# HIMALAYAN MEDICINAL PLANTS FOR THE TREATMENT OF DEPRESSION

*A Source of Rich Antidepressant Agents*

**Cover Photo:**

**Top:** *Eleusine coracana* (L.) Gaertn. (finger millet, also known as ragi in India, Kodo in Nepal). Photo by Dinesh Valke from Thane, India. https://creativecommons.org/licenses/by-sa/2.0, via Wikimedia Commons.

**Middle:** *Hypericum perforatum* (common St. Johnswort). Photo by Agnieszka Kwiecień, Nova. https://creativecommons.org/licenses/by-sa/4.0/deed.en, via Wikimedia Commons.

**Right:** *Mucuna pruriens* (monkey tamarind). Photo by Vijayanrajapuram. https://creativecommons.org/licenses/by-sa/4.0/deed.en, via Wikimedia Commons.

**Left:** *Zanthoxylum armatum* (also called winged prickly ash or rattan pepper in English). Photo by Krzysztof Golik. https://creativecommons.org/licenses/by-sa/4.0/deed.en, via Wikimedia Commons.

*Innovations in Plant Science for Better Health: From Soil to Fork*

# HIMALAYAN MEDICINAL PLANTS FOR THE TREATMENT OF DEPRESSION

## *A Source of Rich Antidepressant Agents*


*Edited by*

**Abdur Rauf, PhD**

**Prabhakar Semwal, PhD**

**Hafiz Ansar Rasul Suleria, PhD**



AAP | APPLE ACADEMIC PRESS

First edition published 2025

**Apple Academic Press Inc.**
1265 Goldenrod Circle, NE,
Palm Bay, FL 32905 USA

760 Laurentian Drive, Unit 19,
Burlington, ON L7N 0A4, CANADA

**CRC Press**
2385 NW Executive Center Drive,
Suite 320, Boca Raton FL 33431

4 Park Square, Milton Park,
Abingdon, Oxon, OX14 4RN UK



*Apple Academic Press exclusively co-publishes with CRC Press, an imprint of Taylor & Francis Group, LLC*



**Library and Archives Canada Cataloguing in Publication**

CIP data on file with Canada Library and Archives

**Library of Congress Cataloging-in-Publication Data**

CIP data on file with US Library of Congress

ISBN: 978-1-77491-630-8 (hbk)
ISBN: 978-1-77491-631-5 (pbk)
ISBN: 978-1-03271-797-5 (ebk)

# Books in Innovations in Plant Science for Better Health: From Soil to Fork series

**Series Editor-in-Chief:**

**Hafiz Ansar Rasul Suleria, PhD**
*School of Agriculture and Food, Faculty of Veterinary and Agricultural Sciences, The University of Melbourne, Parkville, Victoria, Australia*
*Email: h.suleria@hotmail.com*

This book series offers academia, engineers, technologists, and users from different disciplines information to gain knowledge on the breadth and depth of this multifaceted field. The volumes explore the fields of phytochemistry along with its potential and extraction techniques. The volumes discuss the therapeutic perspectives of biochemical compounds in plants and animal and marine sources in an interdisciplinary manner because the field requires knowledge of many areas, including agricultural, food, and chemical engineering; manufacturing technology along with applications from diverse fields like chemistry; herbal drug technology; microbiology; animal husbandry; and food science; etc. There is an urgent need to explore and investigate the innovations, current shortcomings, and future challenges in this growing area of research.

We welcome book proposals on the following specialty areas (but not limited to):

- Food and function
- Nutritional composition of different foods materials
- Functional and nutraceutical perspectives of foods
- Extraction of bioactive molecules
- Phytomedicinal properties of different plants
- Traditional plants used as functional foods
- Phytopharmacology of plants used in metabolic disorders
- Importance of spices and medicinal and functional foods
- Natural products chemistry

- Food processing waste/byproducts: management and utilization
- Herbals as potential bioavailability enhancer and herbal cosmetics
- Phytopharmaceuticals for the delivery of bioactives
- Alternative and complementary medicines
- Ethnopharmacology and ethnomedicine
- Marine phytochemistry
- Marine microbial chemistry

## BOOKS IN THE SERIES:

- **Assessment of Medicinal Plants for Human Health: Phytochemistry, Disease Management, and Novel Applications**
  Editors: Megh R. Goyal, PhD, and Durgesh Nandini Chauhan, MPharm
- **Bioactive Compounds of Medicinal Plants: Properties and Potential for Human Health**
  Editors: Megh R. Goyal, PhD, and Ademola O. Ayeleso
- **Bioactive Compounds from Plant Origin: Extraction, Applications, and Potential Health Claims**
  Editors: Hafiz Ansar Rasul Suleria, PhD, and Colin Barrow, PhD
- **Bioactive Compounds from Multifarious Natural Foods for Human Health**
  Editors: Hafiz Ansar Rasul Suleria, PhD, Megh R. Goyal, PhD, and Huma Bader Ul Ain
- **Cereals and Cereal-Based Foods: Functional Benefits and Technological Advances for Nutrition and Healthcare**
  Editors: Megh R. Goyal, PhD, Kamaljit Kaur, PhD, Jaspreet Kaur, PhD
- **Curative and Preventive Properties of Medicinal Plants: Research on Disease Management and Animal Model Studies**
  Editors: Megh R. Goyal, PhD, Junaid Ahmad Malik, PhD, and Ademola Olabode Ayeleso, PhD
- **Health Benefits of Secondary Phytocompounds from Plant and Marine Sources**
  Editors: Hafiz Ansar Rasul Suleria, PhD, and Megh Goyal, PhD
- **Herbal Medicines and Nutritional Supplements for Health Benefits**
  Editors: Megh R. Goyal, PhD, and Anamika Chauhan, PhD
- **Herbs, Spices, and Medicinal Plants for Human Gastrointestinal Disorders**
  Editors: Megh R. Goyal, PhD, Preeti Birwal, PhD, and Durgesh Nandini Chauhan, MPharm
- **Himalayan Medicinal Plants for the Treatment of Depression: A Source of Rich Antidepressant Agents**
  Editors: Abdur Rauf, PhD, Prabhakar Semwal, PhD,and Hafiz Ansar Rasul Suleria, PhD

- **Human Health Benefits of Plant Bioactive Compounds: Potentials and Prospects**
  Editors: Megh R. Goyal, PhD, and Hafiz Ansar Rasul Suleria, PhD
- **Phytochemical and Pharmacological Investigation of the Family Rutaceae**
  Editors: Abdur Rauf, PhD, Hafiz Ansar Rasul Suleria, PhD, and Syed Muhammad Mukarram Shah, PhD
- **Phytochemicals and Medicinal Plants in Food Design: Strategies and Technologies for Improved Healthcare**
  Editors: Megh R. Goyal, PhD, Preeti Birwal, PhD, and Santosh K. Mishra, PhD
- **Phytochemicals from Medicinal Plants: Scope, Applications, and Potential Health Claims**
  Editors: Hafiz Ansar Rasul Suleria, PhD, Megh R. Goyal, PhD, and Masood Sadiq Butt, PhD
- **Plant- and Marine-Based Phytochemicals for Human Health: Attributes, Potential, and Use**
  Editors: Megh R. Goyal, PhD, and Durgesh Nandini Chauhan, MPharm
- **Plant-Based Bioactive Compounds and Food Ingredients: Encapsulation, Functional, and Safety Aspects**
  Editors: Junaid Ahmad Malik, PhD, Megh R. Goyal, PhD, Preeti Birwal, PhD, and Ritesh B. Watharkar, PhD
- **Plant-Based Functional Foods and Phytochemicals: From Traditional Knowledge to Present Innovation**
  Editors: Megh R. Goyal, PhD, Arijit Nath, PhD, and Hafiz Ansar Rasul Suleria, PhD
- **Plant Secondary Metabolites for Human Health: Extraction of Bioactive Compounds**
  Editors: Megh R. Goyal, PhD, P. P. Joy, PhD, and Hafiz Ansar Rasul Suleria, PhD
- **The Functional Foods: Nutrient and Health Benefits**
  Editors: Megh R. Goyal, PhD, Junaid Ahmad Malik, PhD, and Anu Kumari
- **The Role of Phytoconstitutents in Healthcare: Biocompounds in Medicinal Plants**
  Editors: Megh R. Goyal, PhD, Hafiz Ansar Rasul Suleria, PhD, and Ramasamy Harikrishnan, PhD
- **The Therapeutic Properties of Medicinal Plants: Health-Rejuvenating Bioactive Compounds of Native Flora**
  Editors: Megh R. Goyal, PhD, PE, Hafiz Ansar Rasul Suleria, PhD, Ademola Olabode Ayeleso, PhD, T. Jesse Joel, and Sujogya Kumar Panda

# About the Editors

**Abdur Rauf, PhD**
*Assistant Professor, Department of Chemistry, University of Swabi, Anbar, Khyber Pakhtunkhwa, Pakistan*

Abdur Rauf, PhD, is a distinguished member of the Department of Chemistry at the University of Swabi, KP, Pakistan. Having earned his PhD from the Institute of Chemical Sciences, University of Peshawar, Pakistan in 2015, Dr. Rauf has dedicated his career to advancing the realms of phytochemistry and pharmacology. His primary focus lies in the bio-guided isolation of new compounds for drug discovery. Dr. Rauf's research endeavors encompass the isolation and elucidation of bioactive compounds, employing innovative techniques for in vitro and in vivo biological screening. His scholarly contributions are remarkable, evidenced by his authoring or coauthoring over 430 research papers in esteemed peer-reviewed journals, boasting an impact factor of 1760.65. Garnering over 14,000 citations according to Google Scholar, Dr. Rauf holds an H-index of 60 and an i10 index of 275. Additionally, he has edited five books and contributed 39 chapters to international publications. Notably, he holds one national and one international patent, with four more international patents pending, all pertaining to the discovery of novel antidiabetic drugs.

Dr. Rauf's influence extends beyond academia, as he serves as an associate editor or editorial board member for prestigious journals such as Frontier in Pharmacology, Heliyon, Biocell, Green Processing and Synthesis, and Medicinal Chemistry. He is also a fellow of the American Chemical Society and Bioconlab at Oral State University, Russia. Acknowledged as one of the World's Top 1% Highly Cited Researchers in 2022 by Clarivate, Dr. Rauf's impact is undeniable.

Throughout his career, Dr. Rauf has demonstrated a commitment to nurturing future talent, having supervised three PhD and ten MPhil students. His accolades include the Young Scientist Award from the Directorate of Science and Technology KP, Pakistan in 2018, and the Research Productivity Award for 2016-17 from the Pakistan Science Foundation.

Noteworthy among his achievements is the organization of the 1st International Conference on Drug Discovery against Cancer and other Diseases (DDCD) in 2019, along with participation in the International Pharmacy Conference & Exhibition on Emerging Trends in Drug Development, Therapeutics, and Commercialization (2019).

Driven by a vision for sustainable innovation, Dr. Rauf's future research aims to develop eco-friendly and cost-effective protocols for producing nanoparticle-based drugs with biomedical applications. He also seeks to harness local natural resources, particularly the national flora, to address various health disorders, thereby contributing significantly to the national economy.

Given his exemplary contributions to chemistry and pharmaceutical sciences, Dr. Abdur Rauf is undeniably a deserving candidate for prestigious honors and awards, including civil distinctions such as the Pride of Performance (Pakistan).

**Prabhakar Semwal, PhD**
*Assistant Professor, Department of Life Sciences, Graphic Era Deemed to be University, Dehradun, Uttarakhand, India*

Prabhakar Semwal, PhD, is currently working as Assistant Professor in the Department of Life Sciences, Graphic Era (deemed to be University), Dehradun, Uttarakhand, India. He served as a guest editor for Frontiers in Pharmacology and as a reviewer for *Scientific Reports, Oxidative Medicine and Cellular Longevity, Frontiers in Pharmacology, Ethnobotany Research & Applications,* etc. Before joining GEU, Dr. Semwal worked at the Uttarakhand State Council for Science and Technology, Dehradun (Governement of Uttarakhand) for five years in different capacities. Dr. Semwal also presented papers at several international and national conferences. His major research interests include high-altitude medicinal plants and bioactive compounds and their biological applications. Dr. Semwal has published more than 50 research and review articles in highly reputed journals along with five book chapters. Presently, he is supervising several students and also working on a research project funded by Uttarakhand Council for Biotechnology (UCB), Govt. of Uttarakhand, India.

**Hafiz Ansar Rasul Suleria, PhD**
*School of Agriculture and Food, Faculty of Veterinary and Agricultural Sciences, The University of Melbourne, Parkville, Victoria, Australia*

Hafiz Ansar Rasul Suleria, PhD, is affiliated with the School of Agriculture and Food, Faculty of Veterinary and Agricultural Sciences, The University of Melbourne, Parkville, Victoria, Australia. He was formerly the Alfred Deakin Research Fellow at Deakin University, Victoria, Australia. He did his postdoctoral fellowship at the Department of Food, Nutrition, Dietetic and Health at Kansas State University, USA. He is also an Honorary Fellow of the Diamantina Institute, Faculty of Medicine, The University of Queensland (UQ), Australia. Dr. Suleria has been awarded an International Postgraduate Research Scholarship and the Australian Postgraduate Award for his PhD research at the UQ School of Medicine and the Translational Research Institute in collaboration with the Commonwealth and Scientific and Industrial Research Organization, Australia. Before joining the UQ, he worked as a lecturer in the Department of Food Sciences, Government College University Faisalabad, Pakistan. He also worked as a research associate in the PAK-US Joint Project funded by the Higher Education Commission, Pakistan, and Department of State, USA, with collaboration of the University of Massachusetts, USA, and the National Institute of Food Science and Technology, University of Agriculture Faisalabad, Pakistan. Dr. Suleria has published more than 80 peer-reviewed scientific papers in professional journals and has co-edited several books. He is also in collaboration with more than ten universities where he is working as a co-supervisor/special member for PhD and postgraduate students. His major research focus is on food nutrition, particularly in screening of bioactive molecules, including isolation, purification, and characterization using various cutting-edge techniques from different plants, marine, and animal sources; in vitro, in vivo bioactivities; and cell culture and animal modeling.

# Contents

# Contributors

**Waleed Al Abdulmonem**
Associate Professor, Department of Pathology, College of Medicine Qassim University, Buraydah, Saudi Arabia

**G. Abhilash**
Bangalore Medical College and Research Institute, Bangalore, Karnataka, India

**Zubair Ahmad**
MPhil Scholar, Department of Chemistry University of Swabi, Swabi, Tehsil, Khyber Pakhtunkhwa, Pakistan

**Basiru Olaitan Ajiboye**
Phytomedicine and Molecular Toxicology Research Laboratory, Department of Biochemistry, Federal University Oye-Ekiti, Nigeria

**Muhammad Nadeem Akhtar**
University Institute of Diet and Nutritional Sciences, Faculty of Allied Health Sciences, The University of Lahore, Pakistan

**Mohammad Ali**
Assistant Professor, Center for Biotechnology and Microbiology University of Swat, Swat, Tehsil, Charbagh, Khyber Pakhtunkhwa, Pakistan

**Abdullah S. M. Aljohani**
Associate Professor, Department of Veterinary Medicine, College of Agriculture and Veterinary Medicine, Qassim University, Buraydah, Saudi Arabia

**Jigisha Anand**
Department of Biotechnology, Graphic Era Deemed to be University, Dehradun, Uttarakhand, India

**K. Ashwini**
Department of Microbiology, Biotechnology and Food Technology, Bangalore University, Bangalore, Karnataka, India

**J. P. Shabaaz Begum**
Department of Life Sciences, Graphic Era (Deemed to be University), Dehradun, Uttarakhand, India

**Nadia Bibi**
Assistant Professor, Department of Microbiology, Shaheed Benazir Bhutto Women University, Peshawar, Khyber Pakhtunkhwa, Pakistan

**Ashun Chaudhary**
Department of Plant Science, School of Life Sciences, Central University of Himachal Pradesh, Shahpur, Kangra, Himachal Pradesh, India

**Arti Chauhan**
Department of Plant Science, School of Life Sciences, Central University of Himachal Pradesh, Shahpur, Kangra, Himachal Pradesh, India

**Natália Cruz-Martins**
Faculty of Medicine, University of Porto, Porto, Portugal; Institute for Research and Innovation in Health (i3S), University of Porto, Porto, Portugal; Institute of Research and Advanced Training in Health Sciences and Technologies (CESPU), Rua Central de Gandra, Gandra PRD, Portugal; TOXRUN–Toxicology Research Unit, University Institute of Health Sciences, CESPU, CRL, Gandra PRD, Portugal

**Kiran Dhami**
Department of Life Sciences, Graphic Era Deemed to be University, Dehradun, Uttarakhand, India

**Nisar Ud Din**
MPhil Scholar, Center for Biotechnology and Microbiology University of Swat, Swat, Tehsil, Charbagh, Khyber Pakhtunkhwa, Pakistan

**Muhammad Farooq**
Assistant Professor, Department of Microbiology, University of Swabi, Khyber Pakhtunkhwa, Pakistan

**Anees Ahmed Khalil**
University Institute of Diet and Nutritional Sciences, Faculty of Allied Health Sciences, The University of Lahore, Pakistan

**Wajid Khan**
Assistant Professor, Center for Biotechnology and Microbiology University of Swat, Swat, Tehsil, Charbagh, Khyber Pakhtunkhwa, Pakistan

**Saima Naz**
Institute of Biotechnology and Microbiology, Bacha Khan University, Charsadda, Khyber Pakhtunkhwa, Pakistan

**Leirika Ngangom**
Department of Life Sciences, Graphic Era (Deemed to be University), Dehradun, Uttarakhand, India

**Ahmed Olatunde**
Department of Medical Biochemistry, Abubakar Tafawa Balewa University, Bauchi, Nigeria

**Nishant Rai**
Department of Biotechnology, Graphic Era Deemed to be University, Dehradun, Uttarakhand, India

**Kirti Raina**
Department of Plant Science, School of Life Sciences, Central University of Himachal Pradesh, Shahpur, Kangra, Himachal Pradesh, India

**Abdur Rauf**
Assistant Professor, Department of Chemistry, University of Swabi, Swabi, Anbar, Khyber Pakhtunkhwa, Pakistan

**Muhammad Rizwan**
Assistant Professor, Center for Biotechnology and Microbiology University of Swat, Swat, Tehsil, Charbagh, Khyber Pakhtunkhwa, Pakistan

**Prabhakar Semwal**
Department of Life Sciences, Graphic Era Deemed to be University, Dehradun, Uttarakhand, India

**Rohit Sharma**
Department of Forest Products, College of Forestry, Dr. Yashwant Singh Parmar University of Horticulture and Forestry, Nauni, Solan, Himachal Pradesh, India; Department of Rasa Shastra and

Bhaishajya Kalpana, Faculty of Ayurveda, Institute of Medical Sciences, Banaras Hindu University, Varanasi, Uttar Pradesh, India

**Muhammad Faisal Siddiqui**
Assistant Professor, Department of Microbiology, Hazara University, Mansehra, Pakistan

**Hafiz Ansar Rasul Suleria**
School of Agriculture and Food, Faculty of Veterinary and Agricultural Sciences, The University of Melbourne, Parkville, Victoria, Australia

**Taj-Ud-Din**
MPhil Scholar, Center for Biotechnology and Microbiology University of Swat, Swat, Tehsil, Charbagh, Khyber Pakhtunkhwa, Pakistan

**Sunil Thakur**
Department of Animal Sciences, School of Life Sciences, Central University of Himachal Pradesh, TAB Shahpur, Kangra, Himachal Pradesh, India

**Habibu Tijjani**
Natural Product Research Laboratory, Department of Biochemistry, Bauchi State University, Gadau, Nigeria

**Shikha Tomar**
Department of Life Sciences, Graphic Era (Deemed to be University), Dehradun, Uttarakhand, India

**Mohammad Nazir Uddin**
Assistant Professor, Center for Biotechnology and Microbiology University of Swat, Swat, Tehsil, Charbagh, Khyber Pakhtunkhwa, Pakistan

**Divya Venugopal**
Department of Life Sciences, Graphic Era (Deemed to be University), Dehradun, Uttarakhand, India

# Preface

We introduce this book volume under the book series Innovations in Plant Science for Better Health: From Soil to Fork. This book mainly covers the current research on the importance of bioactive compounds derived from Himalayan plants. This book, Himalayan Medicinal Plants for the Treatment of Depression, is organized into three parts: Part I: Antidepressant Agents from the Himalayan Plants, Vegetables, and Fruits as Antidepressant Agents; Part II: Antidepressant Agents from the Himalayan Flowers, Mushrooms, Grains, Beans, and Nuts; and Part III: Antidepressant Agents from the Himalayan Beverages, Spices, and Microalgae.

In this book, Himalayan plants, vegetables, fruits, flowers, mushrooms, grains, beans, nuts, spices, and beverages, along with their traditional applications, chemical composition, and antidepressant effects, have been discussed with the help of scientific literature.

The main aim of this book is to guide the scientific community on how bioactive compounds from Himalayan botanicals and foods can alleviate various ailments. This volume sheds light on the potential of both foods and plants for human health in different technological aspects, and it contributes to the ocean of knowledge on plant science, food science, and phytotherapy. We hope that this compendium will be useful for students and researchers as well as for persons working in the nutraceuticals, herbal, and food industries.

The contributions by the cooperating authors to this book have been most valuable in the compilation. Their names are mentioned in each chapter, as well as the list of contributors. This book would not have been written without the valuable cooperation of these investigators, many of whom are renowned scientists who have worked in the fields of food science, plant science, and nutrition throughout their professional careers.

We would like to thank Apple Academic Press, Inc., for making every effort to publish this book when all are concerned with health issues. We express our admiration to our families and colleagues for their understanding and collaboration during the preparation of this book volume.

# PART I

# Antidepression Agents From the Himalayan Plants, Vegetables, and Fruits

## CHAPTER 1

# The Himalayan Plants as a Source of Antidepression Agents

NATÁLIA CRUZ-MARTINS[1–4] and PRABHAKAR SEMWAL[5]

[1]*Faculty of Medicine, University of Porto, Porto, Portugal*

[2]*Institute for Research and Innovation in Health (i3S), University of Porto, Porto, Portugal*

[3]*Institute of Research and Advanced Training in Health Sciences and Technologies (CESPU), Rua Central de Gandra, Gandra PRD, Portugal*

[4]*TOXRUN – Toxicology Research Unit, University Institute of Health Sciences, CESPU, CRL, Gandra PRD, Portugal*

[5]*Department of Biotechnology, Graphic Era Deemed to be University, Dehradun, Uttarakhand, India*

## ABSTRACT

Nowadays, depression is considered a serious, life-threatening disorder. Depressive disorders are continuously increasing, affecting people from both developing and developed nations. The main causes responsible for the stimulation of depression include the activation of monoamine oxidase-A and the inhibition of nor adrenaline and 5-hydroxytryptamine (5-HT). Numerous synthetic drugs are available in the market for the treatment of depression; however, some of these drugs are not completely effective, while others have potential side effects. The interest in finding novel plant-based drugs, which are more effective and safer than the

Himalayan Medicinal Plants for the Treatment of Depression: A Source of Rich Antidepressant Agents. Abdur Rauf, Prabhakar Semwal, & Hafiz Ansar Rasul Suleria (Eds.)

currently available ones, has led researchers to conduct intense research. The Himalayan region has long been considered a treasure trove of diverse medicinal flora with pronounced pharmacological activities. Recently, the anti-depressive potential of Himalayan plants and their derivatives has been investigated. This chapter provides an overview of the anti-depressant potential of Himalayan plants and their derivatives, along with pre-clinical and clinical evidence.

## 1.1 INTRODUCTION

Plants from the Himalayan region have been increasingly studied in recent years due to their multiple potentialities [24, 38, 49, 50]. They are characterized by possessing a wide diversity of biomolecules with an extremely diverse chemical composition, which are associated with multiple bioactive effects. Initially, these phytoconstituents were thought to exert mainly protective effects for the plant itself, but they have been increasingly recognized as having health-promoting, disease-preventive, and even therapeutic properties [24, 34, 39, 48, 49, 54]. It is also important to emphasize that, along with this growing interest in studying the multiple potentialities of the flora in the Himalayan region, there has been a growing concern about the ecosystem and the exploitation of existing resources [39, 49, 50, 56]. Therefore, efforts have been made to optimize the use of samples collected from there for research purposes. Equally important is the need to identify the biomolecules responsible for the biological potential of some of the plants in the region, despite the strong evidence supporting the synergistic potential when using the matrix of a plant instead of its isolated constituents [25, 39, 54].

Among the multiple potentialities of existing plants in the Himalayan region, there has been a growing concern in the discovery of matrices with high potential in terms of the treatment of depression [13, 16, 23, 34]. In fact, mental disorders occupy a leading position in terms of the incidence of chronic diseases worldwide. Therefore, several studies have already been carried out to clarify the pathophysiology of the disease, the main risk factors in its genesis, as well as to investigate the mechanisms of action of existing drugs used for other purposes, on drug repurposing. Additionally, and more importantly, to discover new biomolecules isolated from Mother Nature, safer and more effective than the currently available ones for the treatment of these conditions [15].

Recent data published in 2022 by the Global Burden of Disease (GBD) study highlights that mental disorders are currently among the main causes of the burden of disease. The Lancet Commission has emphasized that mental health is a fundamental and essential right to promote the development of all people worldwide [7]. Based on this premise, and considering the data recently published in 2019, in which depressive disorders stand out mostly among females and are the main cause of the highest proportion of disability-adjusted life-years (DALYs) globally, it is crucial to promote greater investment in individuals' mental health [7]. Therefore, considering the current burden of these types of disorders worldwide, it is also extremely important to contribute to the discovery of new drugs capable of ensuring a more effective and safer pharmacological treatment among affected individuals. Additionally, acting at the level of mental health promotion among healthy individuals is essential, either by supplementing the daily diet or by providing new naturally-derived food products capable of helping to address potential deficits that may exist.

In this sense, the use of naturally occurring bioactives, namely derived from fruits, vegetables, flowers, roots and tubers, grains, beans and nuts, mushrooms, microalgae, and spices, among others, for health promotion, disease prevention, and even treatment, is of utmost importance. Thus, the purpose of this chapter is to provide an overview of the antidepressant potential of the Himalayan plants.

## 1.2 HIMALAYAN PLANTS AND DEPRESSIVE DISORDERS: AN UPDATED OVERVIEW

The Himalayan region is composed of a wide variety of plants and other natural products, to which renowned bioactive effects have been attributed. The following subsections briefly focus on the anti-depressive potential of natural products derived from the Himalayan region. A key emphasis is also given to the currently available pre-clinical and clinical evidence related to their ability to prevent, manage, and even treat depressive conditions.

### *1.2.1 ANTI-DEPRESSIVE POTENTIAL OF HIMALAYAN-DERIVED PRODUCTS*

A series of medicinal plants have been reported to exhibit interesting anti-depressive potential (Table 1.1). Asparagus racemosus, an adaptogen used in Ayurveda medicine for treating psychological disorders, has been studied by Singh et al. [51]. They evaluated the anti-depressant activity of *A. racemosus* methanol root extract (MAR) at different doses (100, 200, and 400 mg/kg) in various models. The results from the FST assay showed that MAR reduced immobility, while the LH test indicated an increase in avoidance response at all doses. In the 5-hydroxytryptophan-induced head twitches assay, a concentration of 200 mg/kg of MAR resulted in an increase in the number of head twitches. Additionally, in the clonidine-induced aggression assay, MAR exhibited a facilitatory effect on adrenergic and serotonergic systems by increasing aggressive behavior. However, at a concentration of 100 mg/kg, MAR did not show a significant effect on L-DOPA-induced antagonistic behavior [51].

*Aegle marmelos* is another plant product that play a dynamic part in the prevention of several ailments, including diabetes, diarrhea, cancer, ulcer, fertility, etc., being also already evaluated for its anti-depressant activity. Kothari et al. [27] investigated the antidepressant activity of *A. marmelos* (AM) methanol leaf extract at doses of 75, 150, and 300 mg/kg using the tail suspension test (TST) in an albino mice model. The results showed the highest reduction in immobility at a concentration of 300 mg/kg, while a subeffective dose (75 mg/kg) of AM used in combination with the drugs imipramine and fluoxetine showed significant antidepressant activity. Gosh et al. [18] studied the antidepressant activity of the total sterols fraction and stigmasterol isolated from AM leaves in a mice model using the FST and TST assays. The main findings indicated a significant reduction in the time of immobility by total sterols, stigmasterol, and subeffective stigmasterol combined with fluoxetine in both FST and TST [18]. A further study suggested that the serotonergic and glutamatergic systems might mediate the antidepressant-like effect of total sterols, while only the glutamatergic system mediated the antidepressant activity of stigmasterol [18]. Morus alba (mulberry plant), known for its sericulture importance, is also recognized for its anti-hyperglycemic, anti-hyperlipidemic, anti-inflammatory, cardioprotective, and antidepressant activities. Lim et al. [31] evaluated the antidepressant-like effect of the ethyl acetate fraction derived from *M. alba* root bark in a rat model using the FST. A pronounced antidepressant activity was observed at a dose of 100 mg/kg, with imipramine used as a control. Furthermore, Lim et al. [30]

evaluated the antidepressant-like effect of Sanggenon G (30 mg/kg), a compound isolated from *M. alba* root bark, in a male Sprague Dawley (SD) rat model using the FST. *M. alba* aqueous leaf extract showed significant antidepressant activity when used at a concentration of 200 mg/kg in an albino mice model using the TST and FST [20].

*Withania somnifera*, also known as Ashwagandha, is commonly used in traditional system of medicine, including Ayurveda, Unani, etc., to whom several pharmaceutical properties, like antimicrobial, anti-inflammatory, anti-stress, neuroprotective, antidepressant, analgesic, antitumor, etc., have been attributed [4, 9]. Bhattacharya et al. [4] studied the anti-depressant action of the bioactive compound glycowithanolides (WSG), isolated from *W. somnifera* roots, in a rat model using the swim stress-induced 'behavior despair' test and LH test. The results indicated that WSG significantly possesses anti-depressant activity at concentrations of 20 and 50 mg/kg. Ashwagandha, used in combination with clarified butter and water at a ratio of 1:4:16, was also found to be effective against depression in Swiss albino mice using FST and TST [32]. Antony et al. [1] determined that the aqueous methanol extract of ashwagandha and the withanolide-rich fraction (isolated from *W. somnifera*) exert a remarkable dose-dependent anti-depressant activity in FST.

The anti-depressant potential of *Valeriana officinalis, Valeriana jatamansi*, and *Nardostachys jatamansi* was investigated in terms of acetylcholinesterase (AChE) inhibitory activity and microelectrode arrays by Cornara et al. [8]. Among these three species, *N. jatamansi* possesses the maximum inhibitory activity on AChE (IC50 67.15 μg/mL), followed by *V. officinalis* (IC50 123.30 μg/mL) and *V. jatamansi* (IC50 246.84 μg/mL). The mean firing rate and bursting rate were recorded through a microelectrode array, and a stronger inhibition was recorded for *N. jatamansi* (IC50 18.8 and 11.1 μg/mL), followed by *V. officinalis* (IC50 16.5 and 22.5 μg/mL) and *V. jatamansi* (IC50 68.5 and 89.3 μg/mL) [8].

Other species, including Benincasa hispida, *Bacopa monnieri, Bombax ceiba, Centella asiatica, Curcuma longa, Convolvulus pluricaulis, Crocus sativus, Carthamus tinctorius, Emblica officinalis, Hypericum perforatum, Melissa officinalis, Nardostachys jatamansi, Ocimum sanctum, Mimosa pudica, Nelumbo nucifera, Ocimum basilicum, Tinospora cordifolia, Piper longum, Pyrus pashia, Zingiber officinale, Rhododendron arboreum,* and *Trigonella foenum-graecum*, have also been reported for their interesting anti-depressant activity (Table 1.1).

Taken together, the data presented in this section highlight that the medicinal plants from the Himalayan region possess several other pharmacological activities, including anti-inflammatory, antimicrobial, antitumor, anticancer, antidiabetic, etc. Additionally, these plants also exert promising antidepressant effects. They are continuously being investigated and studied for their antidepressant activities by different research groups. It has been observed that FST and TST are the most commonly used behavioral assays to study the antidepressant effects of plant extracts or bioactive compounds isolated from plants in rat and mice models. Plants from the Himalayan region, namely *A. racemosus, B. monnieri, C. asiatica, C. pluricaulis, C. tinctorious, H. perforatum, M. pudica,* and *T. foenum-graecum* possess significant antidepressant properties. The most often studied chemical components for their antidepressant effects are curcumin (*C. longa*), kaempferol (*C. sativus*), hyperforin (*H. perforatum*), piperine (*P. longum*), 4-hydroxyisoleucine (*T. foenum-graecum*), dehydrozingerone (*Z. officinale*) and 6-Shogaol (Z. officinale). In short, despite the study of several plants and their respective compounds, there are still numerous other plants from the region that should be explored for their anti-depressant activity. Himalayan plant species have great potential.

### *1.2.2 HIMALAYAN PLANTS FOR ANTI-DEPRESSIVE PURPOSES: A SNAPSHOT ON PRE-CLINICAL AND CLINICAL EVIDENCE*

The Himalayan region is rich in terms of fauna and flora, and these features have sparked a growing interest in exploring the region's multiple potentialities. Specifically, regarding the flora in the region and considering the list of plants studied so far with renowned anti-depressant effects, it is worth noting that this represents only a small portion of the overall natural richness in the region. With that in mind, this book is dedicated to addressing the anti-depressant effects of the diverse flora found in the Himalayan region.

Firstly, although the focus of this section is not to provide an exhaustive overview of all plants from the Himalayan region with anti-depressant potential, it is important to emphasize that, from a multiplicity of preclinical studies carried out to date, only 9 clinical studies are registered on ClinicalTrials.gov (Table 1.2). Of these, all of them have already been

**TABLE 1.1** Anti-Depressive Potential of the Himalayan Plants

| SL. No. | Plant Species | Extract/Isolated Component | Test Model | Assays | Concentration Used (mg/kg) | References |
|---|---|---|---|---|---|---|
| 1. | *Benincasa hispida* | Methanolic extract of fruits. | Swiss albino mice. | TST; FST | 100 | [11] |
| 2. | *Bacopa monnieri* | Ethanolic extract of plant (natural and micro-propagated plants). | Mice model | Spontaneous motor activity. | 100 | [53] |
| 3. | *Bombax ceiba* | Ethanolic root extract and its fraction (chloroform, ethyl acetate and n-butanol). | Wistar albino rats. | FST; TST | 200 and 400 | [38] |
| 4. | *Centella asiatica* | Leaves extract | Male Sprague Dawley rats (olfactory bulbectomy model). | Behavioral and physiological parameters during open field activity; behavioral parameters in elevated plus maze. | 3, 10, and 30 | [26] |
| | | Ethanolic leaves extract. | Young adult male Wister rats. | FST | 100 and 200 | [47] |
| 5. | *Curcuma longa* | Curcumin | Male Laca mice | FST | 10, 20, 40, and 80 | [28] |
| 6. | *Convolvulus pluricaulis* | Petroleum ether, chloroform, and ethyl acetate fractions of total ethanolic whole plant extract. | Swiss male albino mice. | FST; TST | 15, 25, 50, and 100 | [14] |
| 7. | *Crocus sativus* | Kaempferol | Male albino mice and Wistar rats. | FST | 100 and 200 | [21] |
| | | Petroleum ether and dichloromethane fractions of aqueous ethanol extract of *C. sativus* corms. | Male ICR species mice. | FST; TST | 150, 300, and 600. | [58] |

**TABLE 1.1** *(Continued)*

| SL. No. | Plant Species | Extract/Isolated Component | Test Model | Assays | Concentration Used (mg/kg) | References |
|---|---|---|---|---|---|---|
| 8. | *Carthamus tinctorius* | Hot water extract. | Male albino mice. | FST; TST | 50 and 150 | [2] |
| 9. | *Cimicifuga racemosa* | Ethyl acetate | Swiss albino mice. | FST; TST | 30, 200, and 400 | [19] |
| 10. | *Emblica officinalis* | Aqueous extract of the fruits. | Swiss young male albino mice. | FST; TST; estimation of MAO-A. | 200 and 400 | [14] |
| 11. | *Hypericum perforatum* | Hydroalcoholic extract and hyperforin. | Male CD-COBS rats. | FST | 1.56, 3.12, and 6.25 (4.5% extract); 3.12, 6.25, and 12.5 (0.5% extract). | [6] |
| 12. | *Mimosa pudica* | Aqueous extract. | Male Wistar rat | FST | 6.0 and 8.0 | [33] |
| | | Ethyl acetate extract. | Adult Swiss albino mice. | FST; TST | 100, 200, and 400 | [42] |
| 13. | *Nelumbo nucifera* | Leaves extract | Male ICR mice. | FST; TST | 200 | [59] |
| | | Fruit extract | Male albino mice. | FST | 50, 100, and 200 | [25] |
| | | Aqueous extract of stem. | Male mice | FST; TST | 200 and 400 | [44] |
| 14. | *Nardostachys jatamansi* | Methanolic extract of rhizomes. | Inbred Swiss albino male mice; Swiss albino male mice. | FST; TST | 200 and 400 | [44] |
| | | Hydro-alcoholic extract of rhizomes. | Swiss albino mice; Wistar albino rats. | FST; TST | 200 and 400 | [40] |

**TABLE 1.1** *(Continued)*

| SL. No. | Plant Species | Extract/Isolated Component | Test Model | Assays | Concentration Used (mg/kg) | References |
|---|---|---|---|---|---|---|
| 15. | *Ocimum sanctum* | Ethanolic extract of leaves. | Male Balb-c mice. | TST | 50, 100, and 150 | [22] |
| | | Aerial part ethanolic extract n-butanol fraction. | Male Swiss albino mice. | FST; TST | 50 and 100 | [37] |
| 16. | *Ocimum basilicum* | Petroleum ether, ethanol, and water extracts. | Albino Wistar rats. | FST; TST | 100, 200, and 400 | [5] |
| | | Aqueous extract | Male Swiss albino mice. | FST | 200 | [55] |
| 17. | *Piper longum* | Piperine | ICR male mice | TST | 1, 3, and 9 | [29] |
| 18. | *Pyrus pashia* | Methanolic extract | Albino rats | FST; Loco motor activity (LMA). | 100 and 200 | [41] |
| 19. | *Tinospora cordifolia* | Petroleum ether extract. | Swiss mice | FST; TST | 50, 100, and 200 | [14] |
| 20. | *Trigonella foenum-graecum* | 4-Hydroxyisoleucine. | Albino Swiss mice. | FST | 10, 30, and 100 | [10] |
| | | Leaves ethanolic extract. | Male Wistar albino rats. | FST; TST | 100, 200, and 400 | [36] |
| | | *Trigonella Foenum-Graecum* seeds flavonoids. | Male KM mice. | FST; TST | 35, 70, and 140 | [57] |
| | | Methanolic seed extract. | Swiss albino mice | FST; TST | 50, 100, and 200 | [3] |
| 21. | *Zingiber officinale* | Hydroalcoholic extract. | Adult male Swiss albino rat. | FST; TST | 150 and 300 | [52] |
| | | Dehydrozingerone. | C57BL/6 male mice. | FST; TST | 100 | [35] |
| | | 6-Shogaol | Male albino mice. | FST; TST | 100 and 300 | [43] |

completed, but results are only available for 1 clinical trial (NCT00048815). Interestingly, when analyzing the plants studied, *Hypericum perforatum* was the most studied (NCT04315597, NCT00451516, NCT00048815, NCT00005013), followed by *Curcuma longa* (NCT01875822) and *Cimicifuga racemosa* (NCT01481025). It is noteworthy that in all clinical studies carried out to date, 2 were of phase 1, 3 were of phase 2, and 2 were of phase 3, with 3 of them not categorizable. In all of these studies, the main objective of the intervention was treatment. Out of the 9 clinical studies conducted, 3 were carried out in Brazil, 2 in the USA, 1 in the United Kingdom, 1 in Puerto Rico, and 1 in Australia, with 1 location being unknown.

Regarding the main objectives of the intervention, the authors aimed to measure the relapse time, remission rate, safety, and tolerability of the given formulation. They also aimed to assess the number of opposing actions, including the incidence of negative and positive symptoms, variation in general psychopathology, cognitive performance and memory improvement, emotional attention, and quality of life of participants.

Looking at the clinical study with available results, it is worth mentioning that in the studied population with minor depression (n=73) from the Massachusetts General Hospital, the subjects (n=26) who received orally 810 mg/day of St. John's Wort (in 3 tablets of 270 mg each) for 12 weeks reported no serious adverse events. When comparing the arm receiving citalopram (20 mg/day, taken orally for 12 weeks) with the arm receiving St. John's Wort, there was no significant difference in the reduction of depressive symptoms severity and safety profiles. This suggests that both groups similarly benefited from the therapeutic intervention received. Unfortunately, no results from other completed clinical trials have been published to date, limiting our ability to draw conclusions.

Nevertheless, considering the data obtained from both *in vitro* and *in vivo* experiments, as well as from this first clinical trial with preliminary results, the Himalayan region may be considered an excellent source of medicinal plants with anti-depressive properties. Promising biomolecules can be further isolated from these plants for the development of safer and more effective drugs for anti-depressive purposes.

**TABLE 1.2.** Clinical Trials Performed Using Himalayan Plants and Their Derivatives for Anti-Depressive Purposes

| NCT Number | Title | Conditions | Interventions | Participants | Phases | Start Year | URL |
|---|---|---|---|---|---|---|---|
| | | | **Interventional** | | | | |
| NCT04315597 | Study to learn about relapse prevention of hypericum extract (a drug based on the plant St. John Wort) in outpatients suffering from moderate depressive episodes. | Moderate depressive episodes (major depression). | Drug: Laif® 900, BAY98-7108<br>Drug: Placebo | 398 adults (18–70 years). | Phase 3 | 2009 | https://ClinicalTrials.gov/show/NCT04315597 |
| NCT02914769 | Antidepressant effects of ayahuasca: a randomized placebo-controlled trial in treatment resistant depression. | Major depression | Drug: Ayahuasca<br>Drug: placebo | 35 adults (18–60 years) | Phases 1–2 | 2014 | https://ClinicalTrials.gov/show/NCT02914769 |
| NCT01875822 | Open-label study of curcumin C-3 complex in schizophrenia. | Schizophrenia, schizoaffective disorder, depression. | Dietary supplement: super-curcumin. | 17 adults (18–65 years). | Phases 1–2 | 2009 | https://ClinicalTrials.gov/show/NCT01875822 |
| NCT01514409 | Sub-chronic effects of 5-hydroxytryptophan on mood and emotional processing in healthy volunteers. | Mood | Dietary supplement: 5-hydroxytrytophane dietary supplement: Pharmaceutical grade silica. | 33 adults (18–65 years). | NA | 2011 | https://ClinicalTrials.gov/show/NCT01514409 |
| NCT01481025 | Efficacy and safety for cimicifuga/hypericum product. | Peri-menopausal disorder. | Drug: Cimicifuga + Hiperico<br>Drug: Cimicifuga Herbarium<br>Drug: Aplause® | 210 females (45–60 years). | Phase 3 | 2011 | https://ClinicalTrials.gov/show/NCT01481025 |

**TABLE 1.2** *(Continued)*

| NCT Number | Title | Conditions | Interventions | Participants | Phases | Start Year | URL |
|---|---|---|---|---|---|---|---|
| NCT00451516 | St. John's Wort and Kava in the treatment of major depressive disorder with comorbid anxiety. | Depressive disorder, major anxiety disorders. | Drug: Herbal medicine (St. John's wort and Kava) | 50 adults (18–65 years). | Phase 2 | 2007 | https://ClinicalTrials.gov/show/NCT00451516 |
| NCT00048815 | Drug therapy to treat minor depression. | Depression | Drug: Citalopram<br>Drug: St. John's Wort<br>Drug: Placebos | 73 adults (18–85 years). | NA | 2003 | https://ClinicalTrials.gov/show/NCT00048815 |
| NCT00005013 | Treatment of major depression with St. John's wort (hypericum) | Major depression. | Drug: Sertraline<br>Drug: Hypericum perforatum (St. John's wort) | Adults (>18 years) | NA | 1998 | https://ClinicalTrials.gov/show/NCT00005013 |
| | | | **Observational** | | | | |
| NCT01283568 | Safety and efficacy study of Gamaline + Hipericin in PMS. | Edema, depression, hot flashes, anxiety, irritability. | NA | 240 females (25–55 years) | NA | 2011 | https://ClinicalTrials.gov/show/NCT01283568 |

*NA*: Not applicable.

## 1.3 SUMMARY

The Himalayan plant species have been studied for their potential in treating depressive disorders, and many of them have shown significant antidepressant effects in the form of extracts or active compounds. However, there are still many species that have not been studied in this context. Therefore, there is immense potential for exploring more plant species with antidepressant activity from the Himalayan region in the future. Currently, these Himalayan medicinal plants have been studied at preliminary, pre-clinical, and clinical levels, although the number of studies is limited. The majority of studies have focused on basic assays, such as behavior assessment using tests like TST, FST, LH, etc., in mice or rat models. Therefore, further research is needed to understand the detailed mechanism of action of these plants and bioactive compounds, as well as their reliability for future clinical use.

## KEYWORDS

- **antidepressant**
- **antidepression agents**
- **depressive disorders**
- **drug**
- **edema**
- **forced swim test**
- **Himalaya**
- **medicinal plants**
- **tail suspension test**

## REFERENCES

1. Alegiry, M. H., El Omri, A., Bayoumi, A. A., Alomar, M. Y., Rather, I. A., & Sabir, J. S., (2022). Antidepressant-like effect of traditional medicinal plant *Carthamus tinctorius* in mice model through neuro-behavioral tests and transcriptomic approach. *Applied Sciences,* 12, 5594.

2. Antony, B., Aravind, A. A., Benny, M., Gupta, N. K., Joseph, B., & Sebastian, A., (2020). Bioactivity guided fractionation and purification of anti-depressant molecule from ashwagandha (*Withania somnifera*). *Current Bioactive Compounds, 16*, 681–686.
3. Assad, T., Siddiqui, N., Rajput, M. A., Sayyar, H. T., Soban, M., & Hameed, T., (2021). Assessment of antidepressant activity of fenugreek seeds methanol extract. *Rawal Medical Journal, 46*, 236–239.
4. Bhattacharya, S., Bhattacharya, A., Sairam, K., & Ghosal, S., (2000). Anxiolytic-antidepressant activity of *Withania somnifera* glycowithanolides: An experimental study. *Phytomedicine, 7*, 463–469.
5. Brar, B., Duhan, J. S., & Rakha, P., (2015). Antidepressant activity of various extract from seed of *Ocimum basilicum* Linn. *International Journal of Science and Research, 4*, 41–43.
6. Cervo, L., Rozio, M., Ekalle-Soppo, C., Guiso, G., Morazzoni, P., & Caccia, S., (2002). Role of hyperforin in the antidepressant-like activity of *Hypericum perforatum* extracts. *Psychopharmacology, 164*, 423–428.
7. Collaborators, G. M. D., (2022). Global, regional, and national burden of 12 mental disorders in 204 countries and territories, 1990–2019: A systematic analysis for the global burden of disease study 2019. *The Lancet Psychiatry, 9*, 137–150.
8. Cornara, L., Ambu, G., Trombetta, D., Denaro, M., Alloisio, S., Frigerio, J., Labra, M., et al., (2020). Comparative and functional screening of three species traditionally used as antidepressants: *Valeriana officinalis* L., *Valeriana jatamansi* Jones ex roxb. and *Nardostachys jatamansi* (D. Don) DC. *Plants, 9*, 994.
9. Dar, N. J., Hamid, A., & Ahmad, M., (2015). Pharmacologic overview of *Withania somnifera*, the Indian Ginseng. *Cellular and Molecular Life Sciences, 72*, 4445–4460.
10. Dhingra, D., & Goyal, P., (2008). Evidences for the involvement of monoaminergic and GABAergic systems in antidepressant-like activity of *Tinospora cordifolia* in mice. *Indian Journal of Pharmaceutical Sciences, 70*, 761.
11. Dhingra, D., & Joshi, P., (2012). Antidepressant-like activity of Benincasa hispida fruits in mice: Possible involvement of monoaminergic and GABAergic systems. *Journal of Pharmacology and Pharmacotherapeutics, 3*, 60–62.
12. Dhingra, D., Joshi, P., Gupta, A., & Chhillar, R., (2012). Possible involvement of monoaminergic neurotransmission in antidepressant-like activity of *Emblica officinalis* fruits in mice. *CNS Neuroscience & Therapeutics, 18*, 419–425.
13. Dhingra, D., & Sharma, A., (2006). A review on antidepressant plants. *Natural Product Radiance, 5*, 144–152.
14. Dhingra, D., & Valecha, R., (2007). Screening for antidepressant-like activity of *Convolvulus pluricaulis* Choisy in mice. *Pharmacol Online, 1*, 262–278.
15. Domínguez, F., González-Trujano, E., Gallardo, J. M., & Orozco-Suárez, S., (2016). Antidepressant medicinal plants and compounds used in traditional medicines in North America. In: Grosso, C., (ed.), *Herbal Medicine in Depression: Traditional Medicine to Innovative Drug Delivery* (pp. 381–431). Springer International Publishing, Cham.
16. Fedorova, Y., Kulpin, P., Kotova, T., Denisova, S., & Beregovykh, G., (2021). Study of the antidepressant properties of some plants, *AIP Conference Proceedings* (p. 050001). AIP Publishing, L. L. C.

17. Gaur, V., Bodhankar, S. L., Mohan, V., & Thakurdesai, P., (2012). Antidepressant-like effect of 4-hydroxyisoleucine from *Trigonella foenum graecum* L. seeds in mice. *Biomedicine & Aging Pathology*, *2*, 121–125.
18. Ghosh, S., Kumar, A., Sachan, N., & Chandra, P., (2022). Evaluation of the antidepressant-like effect of total sterols fraction and stigmasterol isolated from leaves of *Aegle marmelos* and possible mechanism (s) of action involved. *Current Drug Discovery Technologies*, *19*, 26–40.
19. Gowri, N. S., (2019). *Assessment of Anti-Depressant Activity of Anisochilus petraeus Leaves in Swiss Albino Mice*. Master's degree thesis, RVS College of Pharmaceutical Sciences, Coimbatore, India.
20. Halemani, S. S., Acharya, A., Anand, S., Gokul, C., & Adake, P., (2017). Evaluation of the antidepressant activity of aqueous extract of leaves of *Morus alba. The Pharma Innovation*, *6*, 15.
21. Hosseinzadeh, H., Motamedshariaty, V., & Hadizadeh, F., (2007). Antidepressant effect of kaempferol, a constituent of saffron (*Crocus sativus*) petal, in mice and rats. *Pharmacologyonline*, *2*, 367–370.
22. Islamie, R., Ro'is, H., Tristianti, I., Helmi, M., & Wulansari, D. D., (2021). Involvement of monoaminergic system in the antidepressant-like effect of *Ocimum sanctum* (linn.) leaves extract in tail suspension test in mice. *Jurnal Ilmiah Ibnu Sina*, *6*, 143–150.
23. Ismail, H., Amanat, M. A., Iqbal, A., & Mirza, B., (2018). Medicinal plants: A complementary and alternative antidepressant therapy. *Current Pharmaceutical Design*, *24*, 2609–2624.
24. Jamloki, A., Trivedi, V. L., Nautiyal, M., Semwal, P., & Cruz-Martins, N., (2022). Poisonous plants of the Indian Himalaya: An overview. *Metabolites*, *12*, 540.
25. Joshi, R. K., Satyal, P., & Setzer, W. N., (2016). Himalayan aromatic medicinal plants: A review of their ethnopharmacology, volatile phytochemistry, and biological activities. *Medicines*, *3*, 6.
26. Kalshetty, P., Aswar, U., Bodhankar, S., Sinnathambi, A., Mohan, V., & Thakurdesai, P., (2012). Antidepressant effects of standardized extract of *Centella asiatica* L in olfactory bulbectomy model. Biomedicine & Aging Pathology, 2, 48–53.
27. Kothari, S., Minda, M., & Tonpay, S., (2010). Anxiolytic and antidepressant activities of methanol extract of *Aegle marmelos* leaves in mice. *Journal of Pharmacology & Pharmacotherapeutics*, *54*, 318–328.
28. Kulkarni, S. K., Bhutani, M. K., & Bishnoi, M., (2008). Antidepressant activity of curcumin: Involvement of serotonin and dopamine system. *Psychopharmacology*, *201*, 435–442.
29. Lee, S. A., Hong, S. S., Han, X. H., Hwang, J. S., Oh, G. J., Lee, K. S., Lee, M. K., et al., (2005). Piperine from the fruits of piper longum with inhibitory effect on monoamine oxidase and antidepressant-like activity. *Chemical and Pharmaceutical Bulletin*, *53*, 832–835.
30. Lim, D. W., Jung, J. W., Park, J. H., Baek, N. I., Kim, Y. T., Kim, I. H., & Han, D., (2015). Antidepressant-like effects of sanggenon G, isolated from the root bark of *Morus alba*, in rats: Involvement of the serotonergic system. *Biological and Pharmaceutical Bulletin*, b15–00471.

31. Lim, D. W., Kim, Y. T., Park, J. H., Baek, N. I., & Han, D., (2014). Antidepressant-like effects of the ethyl acetate soluble fraction of the root bark of *Morus alba* on the immobility behavior of rats in the forced swim test. *Molecules*, *19*, 7981–7989.
32. MK, J., Prathima, C., Huralikuppi, J., Suresha, R., & Murali, D., (2012). Anti-depressant effects of *Withania somnifera* fat (Ashwagandha Ghrutha) extract in experimental mice. *International Journal of Pharma and Bio Sciences*, *3*, 33–42.
33. Molina, M., Contreras, C., & Tellez-Alcantara, P., (1999). *Mimosa pudica* may possess antidepressant actions in the rat. *Phytomedicine*, *6*, 319–323.
34. Monadi, T., Azadbakht, M., Ahmadi, A., & Chabra, A., (2021). A comprehensive review on the ethnopharmacology, phytochemistry, pharmacology, and toxicology of the mandragora genus; from folk medicine to modern medicine. *Current Pharmaceutical Design*, *27*, 3609–3637.
35. Moorkoth, S., Prathyusha, N. S., Manandhar, S., Xue, Y., Sankhe, R., Pai, K., & Kumar, N., (2021). Antidepressant-like effect of dehydrozingerone from Zingiber officinale by elevating monoamines in brain: In silico and in vivo studies. *Pharmacological Reports*, *73*, 1273–1286.
36. Nayak, R. P., Adake, P., & Hafis, T. K., (2019). Antidepressant activity of *Trigonella foenum* leaves in Wistar albino rats. *International Journal of Basic and Clinical Pharmacology*, *8*, 963–966.
37. Nguyen, T. H., Le, T. X., Pham, T. N. H., Nguyen, M. K., & Kinzo, M., (2021). The antidepressant-like effects of an n-butanol fraction of *Ocimum sanctum* Linn. extract in unpredictable chronic mild stress-induced depression in mice. *Vietnam Journal of Science, Technology and Engineering*, *63*, 55–61.
38. Ojha, A., Arora, R., Khan, R., Sah, S. P., & Upadhyaya, K., (2021). Bioactivity guided fractionation of *Bombax ceiba* L. root extract for antidepressant activity and elucidation of mechanism of action using in silico studies. *Indian Journal of Experimental Biology*, *59*, 539–546.
39. Painuli, S., Semwal, P., Cruz-Martins, N., & Bachheti, R. K., (2021). *Medicinal Plants of Himalayan Forests, Non-Timber Forest Products* (pp. 175–212). Springer.
40. Panchal, R., Goswami, N. N., & Anovadiya, A., (2019). Evaluation of antidepressant activity of hydro-alcoholic extract of rhizomes of *Nardostachys jatamansi* DC per se and in combination with fluoxetine in Wistar albino rats and Swiss albino mice. *International Journal of Basic and Clinical Pharmacology*, *9*, 32–35.
41. Pandey, N., Singh, A., & Pant, J., (2017). Anti-depressant activity of methanolic extract of *Pyrus pashia* leaves in rats. *World Journal of Pharmacy and Pharmaceutical Sciences*, *6*, 1175–1183.
42. Patro, G., Bhattamisra, S. K., & Mohanty, B. K., (2016). Effects of *Mimosa pudica* L. leaves extract on anxiety, depression, and memory. *Avicenna Journal of Phytomedicine*, *6*, 696.
43. Rahangdale, P., Basediya, D. K., Thakur, V., & Dubey, B., (2022). Evaluation of antidepressant effect of 6-shogaol in animal models. *Current Research in Pharmaceutical Sciences*, 46–50.
44. Rahman, H., & Muralidharan, P., (2010). Comparative study of antidepressant activity of methanolic extract of *Nardostachys jatamansi* DC rhizome on normal and sleep deprived mice. *Der Pharmacia Letter*, *2*, 441–449.

45. Rajput, M. A., & Khan, R. A., (2017). Phytochemical screening, acute toxicity, anxiolytic and antidepressant activities of the *Nelumbo nucifera* fruit. *Metabolic Brain Disease*, *32*, 743–749.
46. Rege, N. N., Thatte, U. M., & Dahanukar, S. A., (1999). Adaptogenic properties of six rasayana herbs used in ayurvedic medicine. *Phytotherapy Research: An International Journal Devoted to Pharmacological and Toxicological Evaluation of Natural Product Derivatives*, *13*, 275–291.
47. Selvi, P. T., Kumar, M. S., Rajesh, R., & Kathiravan, T., (2012). Antidepressant activity of ethanolic extract of leaves of *Centella asiatica* Linn. by *in vivo* methods. *Asian Journal of Research in Pharmaceutical Sciences*, *2*, 76–79.
48. Semwal, P., Painuli, S., Cruz-Martins, N., & Thapliyal, A., (2022). Himalayan peony (*Paeonia emodi* Royle) Enlightening bioactive compounds and biological applications towards sustainable use. *Biomolecules from Natural Sources: Advances and Applications*, (pp. 345–361).
49. Semwal, P., Painuli, S., Painuli, K. M., Antika, G., Tumer, T. B., Thapliyal, A., Setzer, W. N., et al., (2021). *Diplazium esculentum* (Retz.) Sw.: Ethnomedicinal, phytochemical, and pharmacological overview of the Himalayan ferns. *Oxidative Medicine and Cellular Longevity*, 2021.
50. Semwal, P., Painuli, S., Tewari, D., Bussmann, R. W., Palni, L. M. S., & Thapliyal, A., (2020). Assesment of non-timber Brahma Kamal (*Saussurea obvallata* (DC.) Edgew.), an important Himalayan. *Ethnobotany Research & Applications*, 19.
51. Singh, G. K., Garabadu, D., Muruganandam, A., Joshi, V. K., & Krishnamurthy, S., (2009). *Antidepressant activity* of Asparagus racemosus in rodent models. *Pharmacology Biochemistry and Behavior*, *91*, 283–290.
52. Singh, R., Jain, R., Mishra, R., & Tiwari, P., (2012). Antidepressant activity of hydroalcoholic extract of *Zingiber officinale*. *International Research Journal of Pharmacy*, *3*, 149–151.
53. Tamboli, F. A., Rangari, V. D., More, H. N., Kutwade, V. H., & Patil, V. M., (2021). Antidepressant effects of natural and micropropagated *Bacopa monnieri* (L.) plant extracts. *Research & Reviews: A Journal of Toxicology*, *11*(2), 17–21.
54. Tewari, D., Atanasov, A. G., Semwal, P., & Wang, D., (2021). Natural products and their applications. *Current Research in Biotechnology*, *3*, 82, 83.
55. Tewari, D., Pandey, H. K., Sah, A. N., Meena, H., Chander, V., Singh, R., & Singh, P., (2015). Phytochemical, antioxidant, and antidepressant evaluation of *Ocimum basilicum, O. tenuiflorum, O. kilimandscharicum* grown in India. *Journal of Biologically Active Products from Nature*, *5*, 120–131.
56. Uniyal, S. K., Awasthi, A., & Rawat, G. S., (2002). Current status and distribution of commercially exploited medicinal and aromatic plants in upper Gori valley, Kumaon Himalaya, Uttaranchal. *Current Science*, 1246–1252.
57. Wang, J., Cheng, C., Xin, C., & Wang, Z., (2019). The antidepressant-like effect of flavonoids from *Trigonella foenum-graecum* seeds in chronic restraint stress mice via modulation of monoamine regulatory pathways. *Molecules*, *24*, 1105.
58. Wang, Y., Han, T., Zhu, Y., Zheng, C. J., Ming, Q. L., Rahman, K., & Qin, L. P., (2010). Antidepressant properties of bioactive fractions from the extract of *Crocus sativus* L. *Journal of Natural Medicines*, *64*, 24–30.

59. Zhao, Y. N., Cao, Y. F., Zhang, Y. H., Lu, Y., Ping, X., Qin, S. K., Liu, S. N., et al., (2020). *Nelumbo nucifera* Gaertn stems (Hegeng) improved depression behavior in CUMS mice by regulating NCAM and GAP-43 expression. *Evidence-Based Complementary and Alternative Medicine, 2020.*

# CHAPTER 2

# Antidepression Agents in the Fruits of the Himalaya

TAJ-UD-DIN,[1] MUHAMMAD RIZWAN,[1] MUHAMMAD FAROOQ,[2] MOHAMMAD NAZIR UDDIN,[1] WAJID KHAN,[1] ZUBAIR AHMAD,[3] and NISAR UD DIN[1]

[1]*Center for Biotechnology and Microbiology University of Swat, Swat, Tehsil, Charbagh, Khyber Pakhtunkhwa, Pakistan*

[2]*Department of Microbiology, University of Swabi, Khyber Pakhtunkhwa, Pakistan*

[3]*Department of Chemistry University of Swabi, Swabi, Tehsil, Khyber Pakhtunkhwa, Pakistan*

## ABSTRACT

People in contemporary civilization suffer from a wide range of psychological problems. Depression and anxiety are the most frequent forms of mental illnesses, and they have a profound influence on both individuals and society. Psychiatric disorders are not treated by fruits, but fruits provide the essential nutrients required for the normal function of the brain and neurons. Fruit cultivation and consumption have a long history, and numerous fruits are mentioned in ancient and holy writings. Fruits are abundant in vitamins, minerals, and a variety of other micronutrients. Advancements in science and technology have demonstrated that fruits contain a variety of biologically active and therapeutically effective components. Fruit consumption provides multiple nutrients essential for the proper functioning of the brain and neurons, helping to suppress

Himalayan Medicinal Plants for the Treatment of Depression: A Source of Rich Antidepressant Agents. Abdur Rauf, Prabhakar Semwal, & Hafiz Ansar Rasul Suleria (Eds.)

symptoms of depression. Several fruits are now being considered for their varied composition and potential health benefits. Due to its location near the equator and various weather conditions, the Himalayan regions are home to a variety of fruits. This chapter provides a thorough explanation of the nutritive value and chemical composition of numerous fruits that have bioactive properties against depression and anxiety.

## 2.1 INTRODUCTION

People in modern civilization experience a variety of psychiatric problems, particularly depression, sadness, anxiety, and sleeplessness. Depressive disorders are well-known types of mental diseases that have a significant impact on individuals and society [62]. According to the World Health Organization (WHO), major depressive disorders (MDDs) will rank as the second most common disease worldwide by 2020 [29]. The global burden of disease (GBD) study from 2010 found that depression was the second-leading cause of infirmity globally and significantly increased the risk of ischemic heart disease and suicide. Since the 1960s, depression has been classified as "major depression" according to the criteria outlined in the Diagnostic and Statistical Manual (DSM). Major depression is characterized by a wide range of symptoms, including anxiety and insomnia [23].

Another common indication of various social mental disorders and other medical conditions is anxiety. In reality, it is a normal human emotion that is closely linked to genuine fear and likely serves as a mechanism for adapting to the psychological environment. Approximately one in five individuals worldwide, at some point in their lives, meet the clinical criteria for an anxiety disorder [23]. Over the past decade, research in psychopharmacology has increasingly focused on the study of anxiety. It has been observed that individuals with anxiety often experience sleep problems as well. The most prevalent sleep disturbance is a subjective complaint of difficulty initiating or maintaining sleep, or that the sleep is of poor quality and insufficient in quantity, and does not provide restoration [86].

Thus, in the complexity of mental health illnesses, sadness, anxiety, and sleeplessness are common comorbid psychiatric conditions [56]. The immunological and cardiovascular systems are compromised as a result of the connection between serious depression, sleeplessness, and anxiety disorders. These psychiatric conditions have an adverse effect on people's daily lives and employment, in addition to lowering their perceived well-being and quality of life [77]. As far as we are aware, the number of individuals deteriorating from mental illness is rising quickly on a global scale. Therefore, finding treatments that are more effective should be a top priority. There are few treatments for psychiatric diseases that are supported by evidence. Currently, the most popular kind of treatment for mood disorders is pharmaceutical therapy. Despite the fact that many medications seem to be important factors in the most severe mental illness instances, there are several concerns that the medications are not always beneficial for all individuals and cause a variety of side effects as well as tolerance (if used for a long time) [46].

Therefore, it is recommended to seek antidepressants that are more effective, start working right away, and have fewer side effects [72]. Numerous research studies have demonstrated that complementary and alternative medicine (CAM), specifically CCAM, is widely used to treat mental issues such as sadness and anxiety. Multiple studies have shown that psychiatric problems, particularly depression and anxiety, often involve the use of CAM. According to a nationally representative study conducted in America, approximately 53.6% of the population with self-reported depression and 4.5% of non-institutionalized adults with insomnia used some form of CAM to treat their psychiatric disorders in the past year. Several academic works have highlighted the value of CAM in the management of mental diseases [28].

The most widely used type of supplementary and alternative medical therapy is herbal medicine [28]. In 2002, 38 million adult Americans, or more than 18.6% of the country's adult population, reported using herbs for health reasons, up from 15% in 1997. Additionally, it has been recently shown that people with psychiatric issues use herbs more frequently than others. In recent years, detailed descriptions of herbs used to treat depression, anxiety, or insomnia have been reported [80].

Various existing treatment/management techniques, such as psychotherapy, electroconvulsive therapy, and medication, have been utilized to alleviate the burden of depression: Psychotherapy employs psychological techniques such as cognitive restructuring, interpersonal counseling, and behavioral activation [84]. For minor depression, psychotherapy can be used alone. However, in cases of mild to severe depression, psychotherapy is combined with antidepressant drugs. Psychotherapy may involve only the depressed individual or may also include others who have a similar illness or whose presence may expedite their recovery [92]. Psychotherapy assists individuals with depression in identifying issues by focusing on reality and the ability to solve difficulties [8].

Yoga and psychoeducational interventions utilizing this type of therapy, as well as educational interventions employing supervised internet-assisted interventions, have all been reported to yield positive results [33]. For severe MDD patients who have not responded to previous treatment modalities, particularly in cases of therapy-resistant depression, electroconvulsive therapy is administered. This treatment involves brief brain stimulation using magnetic or electrical currents, which has shown some degree of success. While the patient is unconscious, electrical induction is performed as an invasive procedure. Electrical stimulation of the subgenual cingulate white matter has been found to reduce symptoms of treatment-resistant depression, potentially by disrupting localized pathogenic activity in limbic-cortical circuits. Other studies have indicated positive outcomes with subcallosal cingulate stimulation using both long and short pulse widths. However, to achieve the greatest benefits, treatment may need to last up to four months or longer. Non-invasive transcranial magnetic and electrical stimulation has also been reported, which can induce changes in cellular activity, impacting connectivity as well as network structure and function [25].

Considering the various theories regarding the origin of depression, these are employed in therapies. Newer methods and a comprehensive treatment plan that addresses all potential etiological hypotheses are needed to prevent or delay the progression of depression.

## 2.2 APPLICATION OF DIET TO THE TREATMENT OF DEPRESSION

Nutrition can be used as a tool in the fight against depression. This is crucial because some theories contend that specific nutrients necessary for healthy brain function also affect the biochemical and neurochemical processes involved in the onset and advancement of depression [71]. Additionally, there is growing data that links dietary habits to the emergence of significant depression, as well as evidence that supports the role of nutrition in preserving mental health [85]. Nutrition is considered to play a part in the synthesis of the monoamines glutamate and γ-aminobutyric acid (GABA), which are linked to depression [24]. Depressed people also have higher levels of the enzyme monoamine oxidase, which is responsible for breaking down the mood-regulating chemicals serotonin, dopamine, and noradrenaline. Because it is commonly seen as appropriate and simple to implement, nutrition can be thought of as a philosophy. It can also be used for health innovations like a natural, secure, and affordable depression cure [24, 80].

## 2.3 NUTRITIONAL DEFICIENCIES AND THEIR ROLE IN DEPRESSED PATIENTS

Fixed oil (omega-3 polyunsaturated) is also known as (ω-3 fatty acids), which are a typical component of cell membranes and are crucial for proper brain function. They play an active role in controlling dopaminergic and membrane-bound enzymes, as well as cellular signal transduction, serotoninergic transmission, and brain eicosanoid production, which is linked to depression. Furthermore, ω-3 fatty acids up-regulate genes that are essential for preserving synaptic function and plasticity, and they also affect the production of neurotrophic substances that regulate emotions, neurogenesis, and cognition [4, 75]. A deficiency in omega-3 fatty acids can impair the functionality of the nervous system, potentially by altering the structure or activity of ion channels, membrane-bound enzymes, and protein receptors. Consuming more omega-3 fatty acids has been associated with improved depression scores and more effective adaptive coping with stressful situations. Additionally, the clinical effects of the eicosatetraenoic acid form of omega-3 fatty acids on depression symptoms have been observed. ω-3 fatty acids may alleviate depression by regulating serotonergic, dopaminergic, and adrenergic transmission, as well as by producing neurotrophic factors necessary for neurogenesis [32].

Ion channels, protein transporters, membrane receptors, and synaptic vesicles are all utilized by brain neurons to communicate with one another. Cognitive impairments and abnormal neurotransmitter levels are caused by deficiencies in proteins. Proteins are large molecules composed of one or more chains of amino acids that serve as the fundamental units and energy supply for cellular development, repair, and reconstruction. There are 22 amino acids, with nine of them considered "essential" and unable to be synthesized independently by humans. Therefore, the primary sources of these essential amino acids are proteins obtained through diet. Rare amino acids such as selenocysteine and pyrrolidine are produced through post-translational modifications of genetically encoded amino acids like lysine and cysteine [69].

Neurotransmitters are derived from amino acids; glutamate serves as the precursor for glutamate and GABA, while tyrosine and tryptophan are precursors for serotonin, dopamine, and noradrenaline, respectively

[6]. Dairy products, poultry, oats, fish, beans, almonds, and whole grains are all rich sources of tryptophan and tyrosine [21]. Tryptophan easily crosses the blood-brain barrier, where it is subsequently converted into serotonin through the enzymatic actions of tryptophan hydroxylase and 5-hydroxytryptophan decarboxylase facilitated by pyridoxal phosphate [36]. The blood-brain barrier (BBB) prevents serotonin from freely passing through, necessitating a continuous intake of tryptophan from food. Inadequate tryptophan diets cause a decrease in serotonin levels in the brain, leading to symptoms of anxiety, obsessions, and compulsions associated with depression. Chronic stress and low levels of omega-3 fatty acids also disrupt serotonin synthesis in the raphe nuclei, resulting in an uncontrolled increase in pro-inflammatory glucocorticoids and cytokines. These uncontrolled increases promote the production of quinolinic acid, a potentially neurotoxic metabolite, by stimulating tryptophan-metabolizing enzymes [19]. Temporarily reducing tryptophan levels in individuals who have recovered from depression can trigger a sudden recurrence of depressive symptoms [88]. Consuming tryptophan helps lower stress levels and cortical hormones, which are associated with depression [58].

Tyrosine is an amino acid that falls into the category of non-essential amino acids. It is also known as 4-hydroxyphenylalanine and can be obtained from food or produced through the hydroxylation of phenylalanine by the enzyme tyrosine hydroxylase. Tyrosine, or L-tyrosine, is readily absorbed by neurons and converted into dopamine and noradrenaline. The first step in this process is the transformation of L-tyrosine into 3,4-dihydroxy-L-phenylalanine (L-DOPA) by the enzyme tyrosine hydroxylase. L-DOPA is then decarboxylated into dopamine by the enzyme DOPA decarboxylase. Dopamine is further converted into noradrenaline by the enzymatic action of dopamine-hydroxylase. Additionally, the enzyme phenylethanolamine N-methyltransferase converts S-adenosylmethionine to adrenaline by attaching a methyl group to the nitrogen of noradrenaline. Distinct neurons in the ventral locus coeruleus, substantia nigra, tegmental nucleus, and hypothalamus are involved in these processes [14, 16, 73]. Deficiency of tyrosine or anomalies in the enzyme tyrosine hydroxylase can lead to a decrease in dopamine levels, which is associated with anhedonia and lack of desire observed in depression.

Alterations in dopamine transporter binding density may also contribute to decreased dopamine levels [15, 76]. A deficit in tyrosine may also result in noradrenaline deficiency, which is linked to decreased alertness, issues with inattention, concentration, and cognitive ability. Tyrosine is a dietary amino acid that enhances cognitive function in both acute and chronic stress models [60]. This diminished stress effect may be caused by the increased synthesis of noradrenaline [91]. Another amino acid obtained through nutrition is glutamate, which the body mostly produces from glucose via alpha-ketoglutarate during the tricarboxylic acid cycle. Foods high in glutamate include dairy products, mushrooms, fish, and vegetables. Although there is some debate regarding the regulation of depression by vitamins, they are reportedly important in its etiology. Several studies show that people with depression consume fewer vitamins and minerals [52, 75]. Water-soluble vitamins include B and C, whereas fat-soluble vitamins include retinol, calciferol, E, and K. Vitamins A, C, and E are three of the body's most important non-enzymatic free radical scavengers. They serve as the body's line of defense against oxidative stress and mental diseases like anxiety disorders and depression [26, 66].

Supplementing with high doses of antioxidants has been proven to reduce the development of vascular disease and neural damage, suggesting that it may be helpful in preventing or treating serious depressive disorders [59]. Vitamin A can be found in cod liver oil, liver, eggs, milk, broccoli, sweet potatoes, carrots, butter, tomatoes, peppers, and pawpaw [7]. Additionally, it is necessary for immune system and vision maintenance, as well as growth, development, and cognition. Although the cause of vitamin A deficiency in depression is unknown, supplementing with it reduces the symptoms of depression [7]. Vitamin C is found in citrus fruits, strawberries, tomatoes, and strawberries. Additionally, this vitamin contributes to the creation of dopamine and noradrenaline, two monoamine neurotransmitters [35]. By reducing oxidative damage, short-term vitamin C supplementation substantially reduces anxiety levels in diabetic patients and reverses depressed symptoms [10, 64]. Among other fruits, goose meat, shellfish, wheat germ oil, sunflower, almond, peanut, avocado, and kiwi are high in vitamin E. It also maintains the composition and operation of the nervous system [21]. Deficits in glutamate and GABA during neurogenesis are associated with depressive symptoms, due to alterations in the enzymes responsible for their metabolism. Consuming vitamin E lowers depressed symptoms [67].

The B vitamins are crucial for the synthesis of monoamine oxidase, DNA synthesis, membrane preservation, and neuronal function. Salmon, yogurt, eggs, dairy, legumes, liver, and fortified cereals are among the abundant foods that contain them. Taking a vitamin B complex nutritional supplement has been shown to significantly reduce the severity of depressed and anxious symptoms over time [53]. Thiamine, biotin, niacin, folate, pantothenic acid, pyridoxine, cobalamin, and riboflavin are all components of the vitamin B complex. Their absence may be a factor in depression, dementia, and cognitive decline [8, 9, 34]. Vitamin B1 or thiamine is crucial for neural function, especially in the creation of energy, anti-nociception, and cognitive ability. Individuals lacking it have been found to have lower levels of GABA, 2-oxoglutarate, and glutamate in their brains [96]. In addition to these, deficiencies in other nutrients are also responsible for depression and psychiatric disorders.

## 2.4 THE USE OF FRUITS TO TREAT DEPRESSION

The utilization of herbs, like nutraceuticals, as well as their metabolites having bioactivity, is another area of investigation. Depression can be treated using fruits or their bioactive byproducts that have psychotropic effects [5]. Fruits contain phytochemicals with a variety of properties that are crucial for maintaining the nervous system [87]. These phytochemicals typically aid in the repair of damaged cells, the reduction of brain cell inflammation, the restoration of neurotransmitter balance, and the recovery of cognitive deficits [70]. The polyphenols are an essential phytochemical with a significant antidepressant impact. Some of them are talked about in relation to how they affect depressed symptoms [22]. This improvement may have been caused by activating the antioxidant pathway. Quercetin, a polyphenolic flavonoid that possesses antioxidant properties and inhibits depressive-like behaviors, is found in varieties of fruits and vegetables [63].

## 2.5 ANTIDEPRESSANT AGENTS IN HIMALAYAN FRUITS

Himalayan region is rich in flora due to its weather, altitude, and fertile soil [94]. Some of the Himalayan's fruits having antidepressant properties are as in subsections.

### *2.5.1 APPLES (PYRUS MALUS)*

The apple is the most significant temperate fruit both economically and commercially [41]. The highest quality apple production occurs in the Jammu and Kashmir region of the Himalayas. The fruit's complex nutritional makeup and phytochemistry are responsible for its enormous health benefits. Due to its involvement in various biochemical processes that promote human growth, development, and overall well-being, it is often referred to as a "wonder fruit." Apples are cholesterol-free but rich in quercetin, flavonoids, anthocyanins, tannins, dihydrochalcones, catechin, and dietary fiber, particularly pectin. Consuming apples helps prevent diabetes and repair oxidative damage to nerve cells [39, 49].

Apples contain numerous macronutrients and polyphenols that have powerful therapeutic effects, including reducing cholesterol, hypoglycemic action, chemoprevention, cardio-protection, and many others. In addition to being calorie-free, apples also aid in preventing constipation, managing obesity, treating depression, and improving dental health [39, 41] (Figures 2.1 and 2.2).

**FIGURE 2.1** Apple fruit.

Apple consumption has been linked to neurological illness and adverse aging effects. Memory loss and Alzheimer's disease, in particular, are associated with acetylcholine and cholinergic receptors [93]. Apple-based products may have the ability to inhibit the synthesis of presenilin 1 protein, which is a key factor in the development of amyloid-b protein, the most common symptom of Alzheimer's disease [41]. Apples also aid in preventing constipation, combating depression, managing weight, and promoting oral health [17].

**FIGURE 2.2** Apple tree.

#### *2.5.1.1 NUTRITIONAL PROFILE OF APPLE*

Apples are a good source of pectin and are categorized as sub-acid foods. Despite having a small energy value, fresh apples are frequently referred to as "wonder fruits" due to the participation of their numerous constituents in various biochemical processes that promote human health. Apples are low in cholesterol and high in dietary fiber [68]. Apples include a significant amount of vitamin C and potassium, but they do not have the same concentration of minerals and vitamins as other fruits [55]. Below is an explanation of the makeup of an apple in more depth.

The primary ingredients of apples are carbohydrates, specifically starch, hemicellulose, cellulose, and pectin. Starch breaks down when the fruit is ripening, producing sugars, which in turn are in charge of synthesizing a variety of flavor precursors. Fresh mature apples contain 10–13% sugars and nearly 15% starch. The three main sugars are glucose (1–4%), sucrose (1–5%), and fructose (3–10%). Apples, with a glycemic index (GI) score of 38, are nevertheless categorized as low GI foods despite containing 90% carbs because fructose is the primary sugar found in apples. Apples possess the least amount of protein of any fruit, which also suggests that there aren't many amino acids present. Glutamic acid and aspartic acid are the two most abundant amino acids in apples, followed by lysine and leucine [3, 57].

Apples have very little fat and are also cholesterol-free. Additionally, the majority of the triglycerides here are unsaturated. Apples are rich in stearic, palmitic, caproic, and oleic fatty acids, with the majority found in the fruit's peel [11]. Apples are a wonderful source of fiber, which supports gut health by retaining water in the colon and promoting bowel movements [65]. Pectin, a highly soluble fiber, acts as a prebiotic and has various positive impacts on gut flora. In apples, pectin is present in amounts between 1% and 1.5% [51].

Apples are one of the main plants that naturally contain phytochemicals. Phytochemicals are essential for a plant's metabolism and defense. They fall into three categories based on their metabolic origins: terpenoids, alkaloids, flavonoids, and related phenolic and polyphenolic substances [13]. Apples are a great source of polyphenols, which include flavonoids. Polyphenols have at least one phenol ring and possess anti-ultraviolet (UV) radiation and anti-pathogen properties. The peel of the apple contains a higher concentration of phenolic chemicals compared to the flesh, resulting in greater antioxidant activity. Examples of polyphenols found in apples include quercetin-3-galactoside and quercetin-3-glucoside [2, 13].

Dihydrochalcone is one of the flavanols found in apples. About 80% of apple polyphenols are flavanols, followed by hydro-cinnamic acid and chlorogenic acid. Apples have the second-highest concentration of phenolic compounds overall and may also contain the highest amount of free phenolic compounds compared to other fruits. Tannins, quercetin, catechin, and dihydrochalcones are the main bioactive compounds found in the fruit, in that order. When investigating the bioavailability of the main phytonutrients in apples, it was discovered that their distribution, metabolism, and absorption in the body, as well as the amount consumed, heavily influence the bioavailability of these phytochemicals. The phytochemicals, including polyphenols, flavonoids, flavanols, and carotenoids, are responsible for the fruit's beauty as well as its reported health benefits. Flavonoids and chlorogenic acid, a derivative of hydro-cinnamic acid, work together to control fruit quality. The red color of the fruit comes from the flavonoid cyanidin-3-galactoside, which is found in the vacuoles of skin cells. On the other hand, chlorogenic acid is responsible for astringency and is the main factor in fruit juice browning due to oxidation [89].

Micronutrients like vitamins and minerals are necessary for the body to function normally [30]. They participate in a number of biological processes as cofactors. Ascorbic acid (vitamin C), pyridoxine, and niacin are the main vitamins (Vitamin B6) found in apples [50]. Apples contain a strong source of lutein and beta-carotene, two different vitamin A precursors [48]. Mineral content is often low in apples. However, some minerals are found in significant quantities. Examples include potassium, phosphorus, and zinc. Among all the minerals found in apples, potassium is the most prevalent element [31].

### 2.5.1.2 PEAR (PYRUS COMMUNIS)

In terms of production and consumption, pears trail apples. Ascorbic acid, iron, and retinol are all abundant in the fruit. The fruit's high fructose content causes a moderate rise in blood glucose levels after consumption. The fruit is more nutritious than others since it contains nearly every amino acid. Due to the presence of numerous phytochemicals, pears are said to have anti-hyperglycemic, antioxidant, and anti-carcinogenic actions [38] (Figures 2.3 and 2.4).

**FIGURE 2.3** Pear fruits.

**FIGURE 2.4** Pear fruits on pear tree.

*Source*: Photo by Alborzagros. https://creativecommons.org/licenses/by-sa/3.0/

### *2.5.1.3 NUTRITIONAL PROFILE*

The nutritional profile of pears is very similar to that of apples. Pears are a great source of fiber, minerals including calcium, phosphorus, and iron, and vitamins like A and C. When comparing pears to other regularly consumed fruits, they contain more fiber overall. In fact, pears have the highest level of insoluble fiber among frequently consumed fruits. Pears also contain a variety of flavonoids (polyphenols), including anthocyanidins, flavan-3-ols, flavanones, flavones, and flavanols. These phytonutrients have been extensively researched and are abundant in pears. Some specific flavonoids found in pears include epicatechin from the flavan-3-ol family and cyanidin from the anthocyanidin category, which is particularly present in red-skinned pears. However, when compared to apples and other fruits, the concentration of flavonoids in pears does not particularly stand out [83].

Pears are considered a low glycemic index (GI) food, with a GI score of 38. They also have slightly fewer carbs than apples. The GI measures how carbohydrates in food impact blood sugar levels. Low GI foods slow down the release of carbs in the body, which helps regulate glucose metabolism, digestion, and absorption, resulting in a gradual increase in blood sugar and insulin levels. Sugar makes up about 60% of the carbs in pears, with an average of 10.4 g of sugar per 100 g of pear fruit. The majority of the sugar in pears is fructose. Pears differ from other fruits due to their high concentration of the sugar alcohol sorbitol. European pears, in particular, have higher levels of sorbitol compared to Asian pears. On average, pear fruit contains 0.3 grams of protein per 100 grams of fresh fruit. The amino acid score of pears is 29, indicating that they contain almost all of the essential amino acids. The main amino acids found in pears are glutamic acid and aspartic acid. While lysine and other sulfur-containing amino acids are not present in plant protein, select pear cultivars do contain lysine [27, 40].

There is very little fat in pears. Pears typically contain 0.1 g of fat per 100 g. The fruit's flesh typically contains a lot of fat, most of which is unsaturated. With an average amount of 43 mg/100 g, linolenic acid (LA) predominates among the fatty acids found in pears [12, 40]. Organic acids are chemical compounds that primarily consist of C, H, and O and are distinguished by the occurrence of the carboxylic and $SO_2OH$ functional groups, which give these compounds their acidic properties. The organic acids are mostly weaker acids because they do not ionize completely in water. Carboxylic, sulfonic, phenolic, and H3PO4 are examples of common organic acids. The strongest of the bunch is phosphoric acid and sulfonic acid, while phenolic acids are considerably weak acids. They work jointly or separately to enhance the flavor of fruits [82]. Pears comprise many organic acids such as ferulic acid, chlorogenic, citric, malic, tartaric, and tartaric. Malic acid is the most common of them. Due to its function in controlling the metabolism of lipids and glucose, chlorogenic acid is of interest [1].

Except for vitamin D, pears are a rich source of all the fat-soluble vitamins. In addition to these, pears also contain vitamin C and B complex vitamins like B9 and B12 [74]. Pears contain significant amounts of all nine important minerals. These components are crucial for supporting not just the regular biochemical processes in plants but also those occurring in the human body. While zinc encourages the formation of immune cells, potassium helps maintain the salinity balance of cells and tissues [18].

Plants generate phytochemicals, which are biological compounds that have been shown to have a positive impact on human health. They do not cause any disorders, unlike vitamins and minerals. Pears are a rich source of phytonutrients, including phenolic acids and anthocyanidins such as flavonoids, cyanidin, and epicatechin, which belong to the flavan-3-ol group. These phenolic chemicals are abundant in the peel of pears. The main phenolic substances in pears are arbutin and chlorogenic acid. Arbutin (hydroxyquinones), a major flavonoid in pears, is followed by secondary flavonoids such as catechin, quercetin, and epicatechin [54]. Carotenoids, naturally occurring fat-soluble pigments, give pears their yellow color. In addition to serving as a coloring agent, they also play a part in a number of biochemical processes, such as boosting retinol content (by acting as the precursor), cell proliferation, cell-cell connection, and immunological function. Carotene, cryptoxanthin, lutein, and zeaxanthin are the most prevalent carotenoids found in pear fruit [37].

### 2.5.2 QUINCE (CYDONIA OBLONGA)

It is a sub-Himalayan fruit crop with enormous commercial and medicinal potential. Only Himachal Pradesh and Jammu and Kashmir can produce it in India [81]. Quince fruit has a fragrant, acidic, or sour flavor and is often pear-shaped and golden yellow in color. The low-fat fruit is abundant in a wide range of necessary dietary fiber nutrients, antioxidants, and minerals that have significant positive effects on health. Due to its tougher texture and flavor, quince is typically boiled before eating or processed into a variety of goods. Quince jam, popularly called "Murrabe Bihi," is a well-known cultivated fruit, along with other processed goods like jelly and confectionery. Quince seed mucilage is a promising hydrocolloid that can be employed in food items as a thickener and bulking agent. Quince has been traditionally used to cure various conditions, such as pneumonia, digestive disorders, chest problems, and sore throats. Additionally, a number of fruit extracts have shown promise as anti-diabetic, antimicrobial, anti-cancer, and anti-inflammatory agents [78, 90] (Figures 2.5 and 2.6).

**FIGURE 2.5** Quince fruits.

**FIGURE 2.6** Quince fruits on tree.

*Source*: Photo by Spone. https://creativecommons.org/licenses/by-sa/3.0/deed.fr

### *2.5.2.1 NUTRITIONAL PROFILE*

One of the most nutrient-dense but often overlooked fruits, quince has a significant positive impact on human health. Due to its rich composition, it is an excellent dietary choice. The fruit contains calories, carbohydrates, fats, and proteins, and is high in vitamin C, pectins (fibers), and minerals [78]. Fresh quince has a carbohydrate value of 15.3 g per 100 g of fresh fruit. In the fruit, sugars make up the majority of the carbohydrates. The fruit has a total sugar content ranging from 9% to 5% and a reducing sugar concentration of 5% to 9%. As a result, quince has a very low percentage of non-reducing sugar. The fruit contains monosaccharides such as D-glucose, rhamnose, L-arabinose, galactose, and mannose. The three major sugars found in quince are fructose, glucose, and sucrose. Quince, like other pome fruits, is low in proteins. The protein value of fresh quince is approximately 0.4 g/100 g. Quince fruit contains 21 free amino acids [78].

Quince is considered a fruit that is low in calories and fat, and it is said to be more nutritious than apples. Quince has 0.1 g of fat per 100 g [95]. The two essential fatty acids found in quince are linoleic acid and oleic acid. A 100 grams of fruit pulp contains 1.6 grams of crude fiber and 1.9 grams of dietary fiber [20]. The majority of the fiber is made up of galacturonic polymers (pectin). Quince has about 1.8 g of pectin. When it comes to mineral content, quince is preferred over apples. According to reports, quince has twice the amount of certain key minerals (sodium, potassium, calcium, magnesium, iron, and phosphorus) compared to apples. Potassium (K), phosphorus (P), and calcium (Ca) are the three main minerals found in quince [43]. The fruit has a high concentration of ascorbic acid (vitamin C). Quince has twice the amount of vitamin C compared to apples. Vitamins A, K, and E are also abundantly present. The primary organic acid in quince is malic acid, which gives

Many metabolites, including phenolic chemicals, terpenes, various volatiles, and organic compounds, are known to be abundant in quince fruit. The range of quince's total phenolic content is 0.2-1.7 g/kg. The majority of phenolic compounds are caffeoylquinic acids (99%), with 5-O-caffeoylquinic acid being the most prevalent (57%) of them [79]. It contains high levels of hydroxycinnamic derivatives, primarily distinguished by 3- and 5-caffeoylquinic acids and polymeric procyanidins. The golden fruit's pulp is mainly composed of 3-O, 5-O-dicaffeoylquinic acid. Quince has been found to contain over 160 identified volatile chemicals, including hydrocarbons, esters, alcohols, carbonyl compounds, lactones, monoterpenes, and nor-isoprenoids. Quercetin-3-O-galactoside, procyanidin, epicatechin, and quercetin-3-O-robinoside are all concentrated in quince pulp [98].

### *2.5.3 CHERRIES (PRUNUS CERASUS)*

The temperate fruit crop known as cherries has a short shelf life and high commercial value [97]. They are a type of low-calorie fruit that is rich in fiber, polyphenols, carotenoids, ascorbic acid, and potassium, among other nutrients and bioactive compounds. Additionally, they are a good source of melatonin, serotonin, and tryptophan. Consumption of cherries is associated with a decline in a number of chronic inflammatory disorders, including diabetes, cancer, cardiac arrest, and arthritis. Furthermore, eating cherries has been proven to enhance sleep, cognitive function, and pain relief following intense activity [61] (Figures 2.7 and 2.8).

**FIGURE 2.7** Cherries fruits.

**FIGURE 2.8** Cherries fruits on tree.

*Source*: Photo by Diako1971. Source: Photo by Diako1971. https://creativecommons.org/licenses/by-sa/3.0/

### 2.5.3.1 NUTRITIONAL PROFILE

Cherries are rich in carotenoids, fiber, polyphenols, vitamin C, and K, among other minerals and bioactive substances. They also contain melatonin, serotonin, and tryptophan, which make up the majority of the macronutrients in cherries. Sweet cherries and sour cherries have different carbohydrate contents, with sweet cherries containing 16 g of carbs per 100 g of the cherry's edible section, and sour cherries containing 12 g per 100 g. Sweet cherries also have 12 g of total sugar per 100 g, with fructose and glucose being the primary sugars, along with sorbitol and maltose. The sugar level in cherries significantly influences their flavor and processing capacity [39, 44].

The main organic acid found in cherries is malic acid, with a concentration of 0.3–0.8 g/100 g of fruit weight in fresh condition. Cherries also contain other acids such as citric, ascorbic, and fumaric acid. The protein content in cherries is approximately 1 g per 100 g for sour cherries and 0.8 to 1.4 g per 100 g for sweet cherries. Cherry also contains tryptophan, an amino acid that contributes to its organoleptic characteristics. The fat content in cherries is generally low, with less than 1 g per 100 g of fruit, and the fats are primarily saturated and cholesterol-free. The lipid fraction of cherries consists of sterol esters, sitosterol, and ursolic acid. The predominant fatty acids in cherry are linolenic, linoleic, oleic, and palmitic acids, which make up approximately 85% of the overall fatty acid profile [39, 44].

Both sweet and sour cherries are great sources of potassium, with 100 g of edible fruit containing roughly 260 and 200 mg of the potassium mineral, respectively. Calcium, phosphorus, magnesium, and salt are other minerals that have been found in cherry cultivars. Cherries have 7–15 mg of vitamin C per 100 grams. Sour cherries are said to contain more vitamin A compared to sweet cherries. Per 100 g, cherries have 2.1 g of dietary fiber. Reductions in blood cholesterol are positively connected with fiber intake. Carotenoids, anthocyanins, flavonoids, phenolic compounds, and flavanols are the main phytochemicals in cherries. Cherries are a rich source of the carotenoids lutein, zeaxanthin, and alpha-carotene. Additionally, flavanols-3-ols, anthocyanins, hydroxycinnamic acid, and hydroxy benzoic acid are the primary components of cherry phenols. These secondary aromatic molecules contribute to flavor synthesis and oxidative stress. The flavones, flavanone, and flavonoids act as antitoxin and antimicrobial agents. The fruit's shiny appearance is due to the anthocyanins, which include cyanidin, pelargonidin, malvidin, etc. Sweet cherries and sour cherries both contain anthocyanidins in amounts ranging from a few milligrams to 80.4 mg per 100 g of fruit [39, 44].

### *2.5.4 FIG (FICUS CARICA)*

A common, healthy, and tasty fruit from the Indian subcontinent is the fig. Its limited and regional cultivation has made it an underappreciated fruit crop. The fig fruit is eye-catching, packed with nutrients, and full of energy. It can be eaten fresh or dried. The flavor of dried figs is chewy and sweet. Figs are cherished as dried fruit because of their distinctive, recognizable flavor. Each serving of figs consumed daily may contribute close to 20% of the required daily intake of dietary fiber. Additionally, the presence of potassium, omega-6, and omega-3 fatty acids in figs aids in controlling cholesterol levels and blood pressure. The presence of several phytochemicals in figs is thought to contribute to their high therapeutic value [39] (Figures 2.9 and 2.10).

**FIGURE 2.9** Fig fruit.

*Source*: Photo by Tijana Drndarski on Unsplash.

**FIGURE 2.10** Fig fruits on tree.

### 2.5.4.1 NUTRITIONAL PROFILE

Numerous religious scriptures, including the Bible and the Quran, mention the fig fruit [47]. The fig is described in the Quran as the softest, gentlest fruit, free of any toxic or indigestible components. Fresh figs are renowned for their delicious flavor and superior nutritional value. They are a good source of fiber, minerals, amino acids, carbs, proteins, and vitamins. Figs also have significant nutritional value in terms of calories, carbs, lipids, phenolics, and enzymes. The nutritional index of the fig fruit is 11, which is greater than that of the apple (9), resin (8), and date (6). Additionally, dried figs are a rich source of nutrients and are frequently included in the diets of the poor in Mediterranean nations [39, 45].

The fig is a fruit high in carbohydrates and dietary fiber. Fresh figs contain 19.2 g of carbohydrates per 100 g of fruit, with approximately 92% of the carbohydrates being sugars. The major sugars in figs are glucose, fructose, and sucrose. Figs have a total sugar content of 16.3 g per 100 g, while dietary fiber makes up only 5 g per 100 g of figs. Certain fig varieties are known to contain D-glucans with significant antibacterial action. Among all edible fruits, figs have the lowest protein content (0.8 g/100 g). However, figs contain the most amino acids of any fruit, including leucine, tryptophan, phenylalanine, lysine, and histidine. Numerous proteolytic enzymes, such as catalase, esterase, lipase, diastase, and peroxidase, are extracted from figs for use in industry. The fruit's latex is also abundant in enzymes like ficin and amylase. Figs have a low fat content of only 0.3 g per 100 g of fresh fruit. Lipid components recovered from fig fruit include triacylglycerols, esterified sterol glycosides, esterified sterols, and phosphatidyl glycerol. Common fatty acids found in figs include myristic acid, stearic acid, oleic acid, and LA. Figs are also a good source of sterols, particularly sitosterol. Fig fruit contains a variety of phytochemicals. Studies on the phytochemistry of fig fruit have shown that it contains a number of bioactive substances. More than 100 bioactive substances, including arabinose, amyrins, carotenes, glycosides, sitosterol, and xanthotoxin, have been found in figs. Additionally, figs are rich in volatile substances such as hydrocarbons and aliphatic alcohols, as well as phenolic compounds, phytosterols, terpenoids, anthocyanins, organic acids, and coumarins. Although figs have a rich phenolic composition, their phenolic concentration is rather moderate. Rutin, epicatechin, gallic acid, catechin, chlorogenic acid, quercetin, retinoid, syringic acid, and ferulic acid are the most prevalent phenolic compounds in figs. Anthocyanin levels are higher in red varieties. In addition, fig fruit contains metallothionein, a unique protein that contains sulfur [39, 45].

Due to its limited production in both human and animal brains, this protein is essential for the healthy operation of the brain. Clearly, the minimum content of figs is similar to that of human milk in terms of minerals. In actuality, it is thought that figs contain more calcium than milk. Iron is the most prevalent mineral in figs. Nearly half as much iron is present in figs as there is in beef liver. Figs also contain large levels of calcium and potassium. Fruits have substantially greater K, Ca, and Fe contents than other regularly consumed fruits. The fruit have less content of sodium and cholesterol aside from that. Figs are the best source of vitamins; large levels of vitamins A and B are found in fig fruits. Thiamine, riboflavin, B5, and B6 are the most common members of the water-soluble group. Some fig types also include vitamin C and vitamin E. Products made from dried figs have reportedly contained vitamin K. Just like other fruits, figs are impacted by the amount of sugar and organic acids present. The fruit has a generally low acid content. Oxalic, quinic, malic, citric, shikimic, and fumaric are the principal organic acids found there [39, 45].

### *2.5.5 APRICOT (PRUNUS ARMENIA)*

The silky apricot, which is indigenous to China, is prized for both its flavor and appearance. The fruit has a low-fat content but a reasonably high protein and carbohydrate content. Due to its active involvement in the production of melanoproteins, selenium, which is known to be present in apricots, is a substance of enormous medical relevance. Due to its many ingredients, apricots contain a number of medicinal properties, including analgesic, anti-asthmatic, laxative, antiseptic, demulcent, emetic, emollient, antipyretic, pectoral, and vulnerary effects [39] (Figures 2.11 and 2.12).

**FIGURE 2.11** Apricot fruit.

*Source*: Photo by Ilona Frey on Unsplash.

**FIGURE 2.12** Apricot fruits on tree.

*Source*: Photo by Elena Mozhvilo on Unsplash.

### 2.5.5.1 NUTRITIONAL PROFILE

Among other fruits of the genus Prunus, apricots are high in carbohydrates and also include a good amount of fiber, minerals, and vitamins. Additionally, they have varieties of bioactive compounds including polyphenols and carotenoids, which play a variety of roles in metabolic processes and effectively fend against oxidative stress. Furthermore, these substances provide plant tissues with colors (red, blue, and purple), greatly enhancing the visual appeal of the fruit. In small numbers, apricots also contain the phenolic acids chlorogenic, coumaric, caffeic, isochlorogenic, p-coumaric, and ferulic. Apricots have a higher carbohydrate content compared to other stone fruits. The total sugar level of fresh apricots is approximately 9 g per 100 g of fruit weight. Similarly, the carbohydrate content ranges from 10 to 13%. The fruit's sugar content consists of sucrose, fructose, glucose, and sorbitol (D-glucitol). However, the main sugar in apricots is sucrose. Apricots also contain 3.73% and 5.71% of reducing and non-reducing sugars, respectively. Malic acid and citric acid are the two organic acids that are most prevalent in apricots.

Other organic acids found in apricots include glyceric, oxalic, fumaric, and lactic. These maintain the stomach's acid-base equilibrium and boost iron's bioavailability. Caffeic, skimmic, and quinic phenolic acids are also present.

A total of 1.5% of the nitrogenous chemicals in apricots are nitrogenous compounds, of which 75% are protein. The fruit has a protein content of 0.67% as a result. The two most common amino acids are glutamic acid and aspartic acid, with traces of amino acids containing sulfur. Apricots are low in fat, just like all other fruits. Apricots provide 0.4 g of fat per 100 g of fruit weight. Triglycerides contain very little saturated fat, with only about 0.2 g of monounsaturated fat. The three main fatty acids in the fruit, palmitic, oleic, and linoleic, are present in quantities of 76 mg per 100 g of fruit. The amount of fiber in apricots is reasonable (11.5%). More than 70% of the total fiber content of apricots is made up of the following: pectin (1%), cellulose, hemicellulose, and pentose. The edible section of the fruit contains 15.5 milligrams of ascorbic acid per 100 grams. Moreover, the fruit is a rich source of water-soluble folate, niacin, choline, and vitamins (vitamins B3, B9, and B12). Additionally, a significant number of vitamins that are fat soluble are vitamins A, E, D, and vitamin K present in apricots. A 100 g serving of fresh fruit provides over 64% of the daily recommended levels of vitamin A [39, 42].

The ash content is used to measure the amount of minerals. Apricots are a significant source of calcium, magnesium, potassium, and phosphorus. Additionally, apricots contain trace levels of selenium (0.40 g), which has gained significant attention due to its potential therapeutic benefits in sialo-protein metabolism. Apricots are rich in carotenoids, which are fat-soluble phytonutrients that act as antioxidants, protecting cells from oxidative damage. Lycopene and carotene are dietary carotenoids found in apricots. Polyphenols, another category of phytonutrients, are widely distributed in plants and are produced as a secondary metabolite to defend against invading organisms. Apricots contain various phenolic components, including hydroxybenzoic acid, hydro-coumarins, hydro-cinnamic acid, proanthocyanins, anthocyanidins, flavanones, flavones, and flavonols. The musky aroma of apricots is attributed to chemicals such as limonene, linalool, terpineol, geranial, and p-cinene [39, 42].

### 2.5.6 GRAPES (*VITIS VINIFERA*)

Since the beginning of time, people have been cultivating grapes, which are now valued as key commercial horticultural products. Grapes are high in vitamin C, retinol, P, and caffeic acid, a potent anticancer compound, but low in fat, Na, and cholesterol. The abundantly occurring resveratrol, a physiologically active and extensively characterized component of grapes, is famous for its many therapeutic benefits. It has strong antioxidant properties. Resveratrol is also said to have potent anti-cancer and chemo-preventive properties. In addition to resveratrol, grapes are also rich in phytoalexin and oleanolic acid, both of which have been linked to favorable health effects [39] (Figures 2.13 and 2.14).

**FIGURE 2.13** Different types of grapes.

*Source*: Photo by Günter Hoffmann on Unsplash

**FIGURE 2.14** Grapes in tree.

*Source*: Photo by Juraj Filipeje on Unsplash.

### 2.5.6.1 NUTRITIONAL PROFILE

The high nutritional value of grapes and their numerous proven health benefits make them well-known. Grapes are abundant in vitamins C and A, as well as potassium. They also contain trace amounts of calcium and phosphorus. In addition, grapes provide other essential minerals such as boron (B), which is necessary for improving bone health. Grapes also contain flavonoids, which give them and grape juice their purple color. Furthermore, grapes are a healthy food for patients with cardiovascular disease (CVD) due to their low sodium, cholesterol, and fat content, as well as their high content of caffeic acid, a potent cancer-fighting compound. Grapes contain non-structural soluble sugars, which play a significant role in the fruit's quality at harvest. The main sugars found in grapes are glucose and fructose, with very small levels of sucrose (less than 2%) also present. As the fruit ripens, glucose and fructose accumulate in equal amounts. Glucose is primarily present in unripe fruit. Sugar synthesis during fruit ripening affects flavor development and the ripeness of the fruit. The sugars in the fruit are digested during wine production and are then transformed into alcohol through ethanolic fermentation. Juice's sugar content determines how much alcohol will ultimately be in the finished product. Around 95% of all soluble solids found in wine are made up of glucose and fructose. In very small amounts (0.24 g/l), non-fermentable sugars like, xylose ribose, rhamnose, and arabinose are referred to as pentoses [39, 99].

Numerous organic acids, such as amino acids, are found in grapes and are crucial indicators of the quality of juice and wine. Tartaric and malic acids make up 90% of the total amount of acid found in fruit. The process of ripening is linked to the production of sugar from starch and the subsequent decrease in acid content. Tartaric acid is more prevalent in mature fruit than malic. Grapes typically contain less than 1% protein, which is made up of peptides and amino acids. By providing nourishment to yeast and LAB, nitrogen-containing chemicals serve as part of the starting material for fermentation and are responsible for wine stability. The amino acid composition of grapes induces the scent of wine through the generation of secondary metabolites. Grapes contain glutamine and glutamate. Wine must contain the highest levels of proline and arginine [39, 99].

Minerals make up 0.2–0.6% of the fruit's overall weight. The most prevalent mineral, potassium (191 g out of 100 g of fresh fruit weight), makes up 50–70% of the juice's total mineral composition. Grapes also contain calcium, sodium, copper, chlorine, iron, manganese, magnesium, and zinc in addition to potassium. An essential antioxidant enzyme called superoxide dismutase (SOD) uses copper and magnesium as cofactors. Iron in fruit is primarily obtained through drying, as in items like raisins. Grapes contain many vitamins, particularly ascorbic acid, in addition to minerals. Red and green variants also contain retinol, phytonadione, carotene, and B complex vitamins [39, 99].

In grapefruit flesh, 0.9 g of dietary fiber is present in 100 g of food. Grape skin contains significant amounts of insoluble dietary fiber, including lignans and uronic acid. Additionally, hemicellulose, which includes the sugar xylose, is a component of insoluble dietary fiber. Grapes contain pectin and protopectin. Grapes typically contain less than 1 g/l of pectin. Pectolytic enzymes have the ability to hydrolyze pectin during fruit ripening, which helps to soften the fruit. Phenolic compounds are the most prevalent phytochemicals in grapes [39, 99].

Phenolic chemicals play a role in maturation reactions, color and flavor development, and browning reactions. The fruit's seeds and skin contain the highest concentration of phenols. Procyanidins, catechins, flavonols, tannins, anthocyanins, and epicatechin gallate are some of the major phytonutrients found in grapes. Anthocyanidins are coloring pigments of various types that give grapes and wines their red and purple hues. Tannins are phenolic chemicals with a large molecular weight that give fruit its color, including yellow, brown, and red, as well as its astringency and bitterness. White grapes are high in flavan-3-ols, while purple grapes contain anthocyanins (catechins). Some important phenolics include ellagic acid, trans-resveratrol, myricetin, and quercetin [39, 99].

Grape skin contains resveratrol at a concentration of 50–100 g/100 g, an essential antioxidant with potent protection against colon cancer and Alzheimer's disease. Grapes include the flavonoids myricetin, quercetin, and 3-O-glucosides, which are potent antioxidants. Additionally, grapes contain a large amount of polyphenols. Grapes contain a significant amount of flavan-3-ols and their derivatives as its main polyphenols. In addition to these, procyanidin, catechins, epicatechin, and epicatechin-3-O-gallate are the other major phenolics found in grapes [39, 99].

### *2.5.7 OLIVE (OLEA EUROPAEA L.)*

The oldest cultivated tree, the olive, has long been revered as a source of healing and a sign of sanctity. Due to their low levels of fiber, protein, and carbohydrates, olives are a fruit low in calories. However, the protein they contain is of good quality as it almost always contains all of the required amino acids. Olives are known to have a variety of medicinal properties, including free radical scavenging, bactericidal activity, chemo-preventive effect, and anti-cancer activity, because they contain certain phenols like chlorogenic acids, oleanolic acid, cholinergic, and maslinic acid. Numerous olive phenols also function as intermediaries in the treatment of Parkinson's, Alzheimer's, and diabetes [39] (Figures 2.15 and 2.16).

**FIGURE 2.15** Olive fruits.

*Source*: Photo by Melina Kiefer on Unsplash.

**FIGURE 2.16** Olive fruits in tree.

### 2.5.7.1 NUTRITIONAL PROFILE

Fruits typically contain 50% water, 5.8% cellulose, 2.2% oil, 19.1% carbohydrates, 1.6% protein, 1.5% inorganics, and 1–3% phenolic chemicals. Olive fruit also contains crucial substances such as pectin, organic acids, and pigments. Organic acids are intermediate products of the synthesis and breakdown of other chemicals that exhibit metabolic activity. Among all edible fruits, olives have the lowest carbohydrate content, with a total sugar content of 0.52%. The main sugar in olives is glucose, followed by fructose. Additionally, olives contain trace levels of the carbocyclic sugar isothiol, which mediates the action of hormones in the brain [39]. Typically, olives have less than 1% protein. However, the protein they contain is of high quality as it almost always contains all of the required amino acids. The two main amino acids found in the fruit are aspartic acid and glutamic acid, respectively. Additionally, some cultivars have been observed to include threonine, valine, arginine, and leucine. Olives also have little fiber. Around 10 olives give 1.5 g of fiber on average. Olives contain 3.3 g of fiber per 100 grams of fruit on average. However, the fiber in the area is of excellent quality and easily absorbed. Typically, cellulose, hemicellulose, and lignin account about 70% of all the fiber in fruit [39].

The richest source of fats is found in olives, the most temperate fruit. Fruit typically contains 15.32 g of oil per 100 g of weight. The pericarp of the fruit contains 96–98% of the oil, whereas the seed has 2–4 g of oil per 100 g. Along with a small amount of saturates and a significant proportion of polyunsaturated (14%), the oil is primarily monounsaturated. The most common fatty acid is oleic acid, palmitic and LA. The percentage of cholesterol in olives ranges between 1.5% and 1.7%. The fruit of the olive contains more than 30 phenolic compounds. Oleuropein (3,4-dihydroxyphenyl ethanol), also known as hydroxytyrosol, and p-hydroxyphenylethanol, are the two primary phenolic alcohols found in olives (tyrosol). The primary phenolic alcohol in olives which are not completely ripened, known as oleuropein, or secoroidoids, is reduced as the fruit ages. Flavonol glycosides and anthocyanins make up the majority of the flavonoid components in olives. Fruits are given sensory and aromatic qualities by the phenolic compounds, which also have medicinal and physiological effects. Different plant components receive antibacterial characteristics from the olive phenols, which are also responsible for the browning of the fruit. Vanillic acid, protocatechuic acid, cinnamic acid, gallic acid, caffeic acid, coumaric, p-hydroxybenzoic acid, and ferulic acid are the main polyphenols found in olives. Chlorophyll gives the fruit its green color, but pheophytin, an intermediary in the breakdown of chlorophyll, gives black olives their color. Zeaxanthin, lutein, and alpha-carotene are the carotenoids found in olives [39].

A fat-soluble vitamin called tocopherol is the main vitamin found in olives. The fruit is said to contain all four forms of tocopherols, with tocopherol having the highest concentration (300 ppm). However, the ripening process has an impact on tocopherol content. The fruit also contains retinol, vitamin B3, and vitamin B6 in addition to vitamin E. Minerals like calcium, iron, potassium, zinc, copper, phosphorous, and selenium are all present in olives. However, calcium, iron, and zinc are the most common among them. Calcium aids in the development of bones, hormone release, and the maintenance of blood clotting enzymes. Olives typically contain 3.3 mg of iron per 100 g of fruit, or about 25% of the RDI [39].

### *2.5.8 POMEGRANATE (PUNICA GRANATUM)*

The pomegranate, which is frequently referred to as a fruit of paradise, is mentioned in numerous holy texts. India is the country that produces the most pomegranates worldwide. Pomegranates are fruits with a high nutritional value, but their hard rind prevents most people from eating them. They are a rich source of protein, selenium, magnesium, vitamin E, fiber, and carbs. The pharmaceutical, pharmacological, and medical effects of pomegranates are attributable to the bioactive molecules of the fruit, such as alkaloids, steroids, tannins, organic acids, triterpenes, and flavonoids, among others [39] (Figures 2.17 and 2.18).

#### *2.5.8.1 NUTRITIONAL PROFILE*

Pomegranate is a fruit that is high in nutrients and low in calories compared to other fruits. It is a source of all the necessary and easily digestible nutrients, which can revitalize those who are mentally or physically exhausted. It is a nutritious diet suitable for young people, the elderly, the ill, and

**FIGURE 2.17** Pomegranate fruit.

**FIGURE 2.18** Pomegranate fruit on tree.

expectant mothers. Pomegranate has the ability to quench thirst and alleviate hunger. It is a relatively good source of carbohydrates, containing about 18.7 g of carbohydrates per 100 g of pomegranate arils, all of which are sugars. Pomegranate is considered a fruit with a high glycemic index (GI). It has a total sugar content of 10 g/100 g, with glucose being the most concentrated (5 mg per 100 g), followed by fructose. Pomegranate also contains a significant amount of sorbitol (4.9 mg per 100 g), which contributes to its laxative effects. Pomegranates are a moderate source of fiber, with an average of 4 g/100 g of fruit. The fiber in pomegranate consists of lignans and indigestible carbohydrates. The main fiber found in the fruit is pectin (1.4%). Pomegranate contains approximately 0.64% soluble dietary fiber and 0.18% insoluble dietary fiber [39].

Pomegranates contain a decent quantity of protein, unlike the majority of fruits. Fruits have an average protein content of 1.67%. The three main amino acids in pomegranates are aspartic acid, arginine, and glutamine, followed by glutamic acid and histidine. Additionally, there are other amino acids present, including valine, tryptophan, isoleucine, alanine, phenylalanine, methionine, and lysine. Trace amounts of leucine and glycine are also found. Pomegranates have a modest fat content of 0.72 g per 100 g of the fruit. The fruit contains 0.19 g of saturated triglycerides and 0.14 g of saturated fats, with saturated triglycerides being predominant. Pomegranate arils contain 0.050 g of omega-3 (linoleic acid) per 100 g and 0.048 g of oleic acid per 100 g [39].

Pomegranates are a great source of vitamin K, which prevents clotting and hemorrhaging, as well as tocopherol (vitamin E), which prevents sterility. Pomegranates also contain significant amounts of B complex vitamins such as folic acid (vitamin B9) and pantothenic acid. Niacin, riboflavin, and thiamine are also present in trace amounts. Pomegranate is relatively low in ascorbic acid compared to most fruits. Magnesium, potassium, phosphorus, calcium, and sodium make up the extensive mineral composition of pomegranate. The fruit is primarily rich in potassium, but it also contains iron, zinc, manganese, and selenium. Selenium is a crucial trace element due to its role as a cofactor in the formation of selenoproteins. Copper is another trace mineral found in pomegranates. Pomegranate fruit contains over 100 identified compounds. Phenolic chemicals such as phenolic acids, phenylpropanoids, and tannins are examples of the simple molecules found in pomegranates (lignans, melanin, tannins). Pomegranates are known for their high tannin content, including hydrolysable tannins such as pumicalagin, peduneulagin, and punicalin. The majority of phenols in pomegranates are flavonoids. Luteolin and quercetin are among the flavonoids found in pomegranates. The skin and peel of pomegranates get their color from the largest subgroup of flavonoids, anthocyanins [39].

### 2.5.9 KIWI (*ACTINIDIA DELICIOSA*)

The enormous medicinal benefits of kiwifruit, also referred to as Chinese gooseberry, have led to its considerable commercial importance in recent years. China is the primary kiwi grower, while India has only recently begun cultivating the fruit. As the state's most nutrient-dense underused crop, locally farmed kiwis require a great deal of research. Actinidin, a crucial amino acid found in kiwis, enhances stomach function and facilitates digestion [39] (Figures 2.19 and 2.20).

Kiwi is also recognized for having a special protein that has been shown to have anti-inflammatory properties, reduce oxidative stress, and improve gastrointestinal physiology. Actinidia plants have historically been used to cure a wide range of ailments, including gastrointestinal issues, rheumatism, dyspepsia, and hemorrhoids [39].

**FIGURE 2.19** Kiwi fruit.

*Source*: Photo by engin akyurt on Unsplash.

**FIGURE 2.20** Kiwi fruit on tree.

### *2.5.9.1 NUTRITIONAL PROFILE*

Because they contain a high amount of water, kiwis are low in nutrients. They also offer various other beneficial chemicals like tocopherol, flavonoids, and some vital elements, in addition to being a great source of ascorbic acid. Fuzzy kiwifruit and golden kiwifruit have significantly different amounts of total sugar (9 g per 100 g and 11 g per 100 g, respectively). In kiwifruit, glucose and fructose make up the bulk of the sugars. The starch content of the fruit decreases while its overall sugar content rises as it ripens. Protein is not abundant in kiwis, with an average of 1.23 g of protein per 100 g of the edible fruit. All the amino acids are present in kiwifruit, although some are only in minimal amounts. The two most common amino acids are aspartic and glutamic [39].

Actinidin, a cysteine protease and another essential amino acid (EAA) found in kiwi, supports healthy stomach function and promotes the optimum digestion of protein. Actinidin, however, is also referred to as a moderate allergen. Kiwifruits have dietary fiber levels comparable to those of high-fiber foods such as apples, oranges, and bananas. The majority of the fruit's fiber content is made up of lignin, hemicelluloses, and celluloses. One-third of the fruit's dietary fiber is soluble, while the remaining two-thirds are insoluble. Green kiwifruit contains more dietary fiber compared to gold kiwi. The fruit's soluble fiber consists of pectin polysaccharides, while the insoluble fiber is primarily composed of lignin, hemicelluloses, and celluloses. The majority of the fruit's lipids are found in the seeds. Fresh kiwifruit contains 0.52 g of fat per 100 grams, with the majority being unsaturated and cholesterol-free. Kiwi is rich in vitamin C, with fuzzy kiwi fruits containing 92.7 mg of ascorbic acid per 100 g of fresh fruit. Kiwifruit also contains Phytonadione, folate, niacin, choline, and retinol in addition to vitamin C. Berries have three times higher vitamin C concentration than oranges and strawberries on an edible flesh weight basis. Additionally, kiwifruit does not turn brown after being cut due to its high vitamin C concentration and low tannin level. The presence of vitamin C enhances the bioavailability of an essential micronutrient called iron. Kiwifruits have an ash content of 0.61 to 0.76 grams per 100 grams, with potassium being the most abundant element followed by calcium, phosphorus, and magnesium. Kiwifruit also contains a small amount of selenium. Among all fruits, kiwifruits have one of the best salt to potassium ratios [39].

### *2.5.10 BER OR JUJUBE (ZIZIPHUS MAURITIANA)*

Ber, sometimes referred to as Chinese dates, is incredibly nutrient-dense but rarely used. The fruit can provide great nutritional value at a low cost because it contains higher amounts of vitamin C, protein, and minerals compared to apples and mangoes, as well as more phosphorus and iron than oranges. Additionally, it is rich in fiber, which helps maintain normal glycemic response and plasma cholesterol levels in the body. The fruit's pulp is also abundant in phytonutrients, making it a good option as a functional food. Ber has been used to treat various disorders due to its rich composition, including its immune-stimulating, gastrointestinal protective, anti-inflammatory, anti-carcinogenic, anti-obesity, antioxidant, and hepatoprotective properties [39] (Figures 2.21 and 2.22).

**FIGURE 2.21** Ber or jujube fruits.

*Source*: Photo by Frank C. Müller. https://creativecommons.org/licenses/by-sa/2.5/

**FIGURE 2.22** Ber or jujube on tree.

*Source*: Photo by Soyoung Han on Unsplash.

#### *2.5.10.1 NUTRITIONAL PROFILE*

Fruits high in nutrients include Ber. Compared to apples and mangoes, it has more protein, minerals, and vitamin C, while oranges contain more phosphorus and iron. Ber is rich in essential vitamins, minerals, fats, proteins, and carbohydrates, including ascorbic acid, vitamin B complex, retinol, calcium, iron, carotene, phosphorus, potassium, and magnesium. The fruit is affordable and provides ample nourishment. The organic acids malic, citric, succinic, and oxalic contribute to the flavor of Ber. To meet the recommended daily allowance of ascorbic acid and vitamin B complex, consume 1–3 Ber fruits each day. Ber also contains a good amount of fiber, which helps maintain normal glycemic response and plasma cholesterol levels in the body. Ber pulp is an excellent choice as a functional food due to its rich phytonutrient content. The fruit has an acidity of 0.12–1.5% and a TSS concentration of 12–20%. The Ber fruit contains the following polysaccharides: fructose (21.6%), xylose (31%), mannose (12.9%), and glucose (23%). Small amounts of sorbitol and rhamnose are also present. Postponing fruit harvesting reportedly increases the total sugar content. Different reducing and non-reducing sugars accumulate to varying degrees depending on the cultivar and ripening stage. Ber pulp has a soluble protein concentration of 1.20 g/100 g, which is said to decrease as ripening progresses. Each fruit possesses 2–10 amino acids, with different cultivars having different amino acid profiles. The most prevalent amino acids in wild samples are alanine, proline, aminobutyric acid, threonine, and glutamine. In cultivated samples, the measured amounts are arginine, proline, cysteine, cystine, and alanine. When the fruit is fully ripe, free amino acids outnumber protein content. The fruit also contains amino acids that contain nitrogen. Ber has a high dietary fiber content, which is associated with numerous health benefits and has satiety-inducing effects. Pectin in the apple is water-soluble and becomes more abundant as the fruit matures. The main monosaccharide in pectin is arabinose, and during ripening, the side branches that contain rhamnose x, galactose, and rhamnose depolymerize. Jujube fruit pectin can be utilized as a thickening agent and gelling agent in a variety of food industry [39].

The major polysaccharides found in jujube include glucose, fructose, xylose, mannose, and hemicellulose. Jujube fruit extracts contain both an acidic polysaccharide fraction and a neutral polysaccharide fraction. The acidic fraction exhibits stronger superoxide anion scavenging activity and metal ion chelating properties. Jujube fruit, also known as Ber, is a good source of essential fatty acids. Unsaturated fatty acids make up a significant portion of the total fat content (65–75% of total fat), with oleic, linoleic, and palmitoleic acids being the main fatty acids found in Ber pulp. The lipid fraction of Ber fruits contains various fatty acids, including linoleic acid and palmitoleic acid, which contribute to the omega-6 fatty acid content. Ber fruit is known to have higher concentrations of calcium, phosphorus, vitamin C, and carotene compared to apples. It also has a higher vitamin C concentration than oranges and can be used as a natural treatment for ascorbic acid insufficiency. Additionally, the fruit contains significant levels of iron, niacin, thiamine, riboflavin, fluorine, magnesium, zinc, and salt. Dried jujube fruit is rich in vitamin B6, niacin, thiamine, and pantothenic acid, which aid in the metabolism of essential macronutrients. Calcium in jujube fruit helps maintain strong bones and teeth, as well as blood coagulation, muscle contraction transmission, and nerve impulse. Dried jujube contains 79 mg of calcium, which is 8% of the recommended daily intake (RDI), making it an excellent source of calcium. Jujube berries are highly antioxidant-active and contain a higher concentration of bioactive phytonutrients compared to cherries, grapes, and apples. The four main phenolics found in jujube are caffeic acid, p-coumaric acid, hydroxybenzoic acid, and ferulic acid. Jujube fruit extracts also contain flavonoids, carotene, tocopherol, anthocyanins, and proanthocyanins. The fruit is also rich in tannins, which possess blood clotting and anti-inflammatory properties [39].

### *2.5.11 PLUM (PRUNUS DOMESTICA)*

The edible plum is a member of the Prunus genus, Rosaceae family. India is the second-largest producer of plums in the world. Despite being low in calories, plums are rich in vitamins, minerals, and phytochemicals. Plums contain various bioactive substances such as carotenoids, anthocyanins, phenolic acids, minerals, and pectin. In Ayurvedic and Unani medicine, plums and their components are commonly used to treat conditions like leucorrhea, irregular menstruation, and miscarriage. Numerous studies have shown that plums are beneficial for health [39] (Figures 2.23 and 2.24).

**FIGURE 2.23** Plum fruits.

**FIGURE 2.24** Plum fruits on tree.

### *2.5.11.1 NUTRITIONAL PROFILE*

Plums have few calories because they are a concentrated source of phytochemicals, minerals, and vitamins. As plums ripen, their mineral content increases. Additionally, plums are rich in bioactive substances such as minerals, anthocyanins, carotenoids, phenolic acids, and pectin.

In terms of nutrition and dietary value, plums are an important part of our diet. About 15 g of carbohydrates per 100 grams of plums are included in the diet, making them a readily available source of energy. In plums, disaccharides (sucrose) and monosaccharides (glucose and sucrose) make up the majority of the carbohydrates and contribute to the fruit's sweetness. The amount of carbohydrates in plums varies depending on the cultivar, level of maturity, and shelf life. Additionally, plums contain sorbitol, a sugar alcohol with a concentration of 2 g per 100 g, which contributes to their laxative properties. Most stone fruits are said to contain pectin, a form of dietary fiber, with plums having the highest concentration at 1% [39].

Shikimic acid, citric acid, and fumaric acid are present in plums, but malic acid predominates among them. During the growth of the fruit, quinic acid predominates, but as it ripens, malic acid becomes more prevalent. Plums contain ascorbic acid as the principal vitamin. They aid in the metabolism of lipids, carbohydrates, and proteins as a minor source of complex vitamins. With only 0.70 g of protein per 100 grams, plums are a poor source of protein. With an amino acid score of 39, plums contain the amino acids lysine, methionine, phenylalanine, and tryptophan [39].

Plums have no cholesterol and only 0.3 g of good fats. Monounsaturated fats make up 0.2 g and polyunsaturated fats make up 0.1 g, respectively. Fatty acids are also present in plums. The fruit's main saturated and monounsaturated fatty acids are butyric and oleic acids, while its PUFAs are linoleic and linolenic. Plums are rich in bioactive phytonutrients such as carotenoids. The main carotenoid found in plums is cryptoxanthin, along with carotene [39].

### *2.5.12 LITCHI (LITCHI CHINENSIS)*

Litchi, a fruit from the soapberry family, is highly cherished among all edible fruits. Botanically classified as a nut, this fruit is rich in amino acids, carbohydrates, minerals, and vitamins. Litchi contains the bioactive polyphenol oligonol, which has been found to have various beneficial effects on chronic conditions. Studies have shown that litchi aglycones can protect brain cells from apoptosis and prevent cancer. Additionally, consuming litchi has been associated with improved bone density and blood production [39] (Figures 2.25 and 2.26).

**FIGURE 2.25** Litchi fruits.

*Source*: Photo by Ivar Leidus. https://creativecommons.org/licenses/by-sa/4.0/

**FIGURE 2.26** Litchi fruits on tree.

*Source*: Photo by Arwin Neil Baichoo on Unsplash.

### 2.5.12.1 NUTRITIONAL PROFILE

Litchi is a delicious fruit that is rich in healthy elements. The fruit is low in fats and proteins but high in carbs and fiber. It also contains a significant amount of vitamin C. In addition to meeting basic nutritional requirements, the fruit contains various nutraceuticals that promote health. Every 100 grams of edible litchi fruit contains about 18 g of carbohydrates. The majority of these carbs are sugars. On a dry weight (DW) basis, the edible fruit component has a total sugar content of 55.9%, with reducing sugars accounting for 41.5%. Litchi mainly contains sucrose, fructose, and glucose as sugars, with traces of rhamnose, arabinose, galactose, and xylose. Galactose is the primary sugar, along with smaller amounts of methyl inositol and myoinositol. The fruit has a low protein content, with about 1 g of protein per 100 g of fresh, edible fruit pulp. Methionine and tryptophan are the two main amino acids in lychee, although other amino acids have also been reported. Litchi is also abundant in aminobutyric acid (GABA), a neurotransmitter that regulates nerve impulses and plays a significant role in our central nervous system (CNS) [39].

During the development of fruit, succinic acid is more prevalent. However, malic acid predominates as the fruit ripens further. As relatively minor acid components, citric, malonic, clavulanic, glutaric, succinic, and lactic acids are present. The fruit contains 1.5 g of fiber-based carbs. Approximately 46% of the total fiber is soluble fiber, while the remaining content is insoluble fiber. Litchi is a rich source of potassium, just like other fruits. Additionally, the fruit contains selenium and nitrogen. In particular, Vitamin B1, riboflavin, vitamins B6, B3, ascorbic acid, and folate are abundant in the fruit. Lychee is an excellent source of ascorbic acid, with concentrations ranging from 27.6 to 71.5 mg/100 g depending on the cultivar. As the total sugar content increases, the concentration of ascorbic acid also grows. However, as a person ages, less acid is produced due to slower sugar accumulation. Litchi is a rare source of lipids and does not contain cholesterol. The fatty acids found in it are necessary for the body to effectively absorb fat-soluble vitamins [39].

Polyphenols are present in moderate levels in lychee. The fruit has a total of 0.47 to 1.6 mg of phenols. Flavan-3-ol monomers and dimers, which make up roughly 87% of the total polyphenol content of lychee, are among the fruit's polyphenolics. Cyanidin-3-glucoside represents around 92% of all anthocyanins. Oligonol, a significant short-chain polyphenol, is found in lychee due to its possible biological effects. The fruit contains comparatively more flavonoids than phenolics. The overall flavonoid and phenolic content of the fruit decreases as it reaches maturity. Additionally, the fruit's pericarp contains flavanol-3-ol, a mixture of oligomerized polyphenols, as well as flavanol monomers, dimers, and trimers, kaempferol, proanthocyanin A1, A2, and rutin [39].

### 2.5.13 AMLA (*EMBLICA OFFICINALIS*)

The most significant fruit crop on the Indian subcontinent, amla, also known as the Indian gooseberry, is prized for its exceptional nutritive and restorative qualities. The fruit is valued for its exceptional medicinal capabilities in Unani and Indian holistic health systems. Amla has a higher concentration of vitamin C than both oranges and tangerines. The amla fruit also contains high concentrations of calcium, phosphorus, iron, and magnesium. Manganese, cobalt, and chromium are a few trace elements in amla that are important for health [39] (Figures 2.27 and 2.28).

**FIGURE 2.27** Amla fruit.

**FIGURE 2.28** Amla fruit on tree.

*Source*: Photo by Dileesh Kumar on Unsplash.

The fruit's juice has historically been used to cure conjunctivitis, asthma, bronchitis, glaucoma, and lung tuberculosis. The fruit is also used to treat ophthalmic disorders, headaches, coughs, skin disorders, leprosy, and gray hair. Moreover, the fruit is utilized as a laxative, diuretic, and antiscorbutic. These properties make it useful for treating colds, gastric problems, scurvy, and diarrhea, as well as diabetes. Additionally, numerous studies have demonstrated the cardioprotective benefits of amla on the heart, lungs, and kidneys [39].

#### 2.5.13.1 NUTRITIONAL PROFILE

The highest source of vitamin C is amla, or Indian gooseberry. It also contains a lot of vitamins and minerals. Calcium is the most widely occurring mineral in amla fruit powder, followed by phosphorus, iron, and magnesium. For individuals, 50 g of amla powder is sufficient to meet the Recommended Daily Allowance (RDA) for calcium. Additionally, it has been discovered that amla contains significant amounts of trace minerals such as manganese, iron, cobalt, and chromium. The primary fatty acids in amla are octadecatrienoic acid, tridecanoic acid, and omega 3 fatty acids, while the main carbohydrate in amla is glucose [39].

The oil extracted from the amla seed, which is around 17% finer, is yellowish in color and contains linoleic acid, oleic acid, and linolenic acid, as well as stearic acid, palmitic acid, and myristic acid. Dodecane, hydroquinone, phthalic acid, and benzoic acid, which are minor components in amla oil, contribute to its antibacterial qualities. Amla's bitterness is caused by a high concentration of ellagitannins. Alkaloids, benzoyl compounds, diterpenes, sterols, flavonoids, and furan lactones, as well as tannins, gallic acid, quercetin, and phloemic chemicals, are the main phytochemical classes identified in amla. Amla has a tart, astringent flavor and contains significant amounts of pectin and ascorbic acid. Due to the presence of two hydrolysable tannins, emblicanin A and emblicanin B, amla has free radical scavenging properties. The fruit is a good source of ascorbic acid as well as minerals like iron, calcium, phosphorus, carotene, thiamine, and riboflavin [39].

Compared to apples, the amla fruit has a protein concentration that is three times higher and an ascorbic acid concentration that is 160 times higher. Phosphatides, an essential oil, and a fixed oil are all present in the fruit's seeds. This tree contains a lot of tannin in its fruits, bark, and leaves. The bark and the root both contain leucodelphinidin and ellagic acid. A stable oil (16%) with a brownish-yellow color is produced by the seeds. Linolenic (8.8%), palmitic (3.0%), myristic (1.0%), stearic (2.15%), linoleic (44.0%), and oleic (28.4%) fatty acids are included in it. The fruit includes a molecule called leucoanthocyanins (Polyphenol), which prevents vitamin C from oxidizing. It is a decent source of nicotinic acid and vitamin B. The presence of numerous ellagitannins gives amla its overall bitter flavor. Additionally, it includes flavonoids, ellagic acid, and gallic acid, as well as punicafolin, phyllanemblinin A, and phyllanemblinin [39].

### *2.5.14 PEACH (PRUNUS PERSICA)*

Peaches, members of the Prunus genus and Rosaceae family, are highly regarded for their unique flavor and vibrant color. While peaches may be lacking in certain nutrients, they are rich in carotenoids such as beta-carotene, alpha-carotene, and cryptoxanthin, which serve as precursors to vitamin A. However, they are relatively low in proteins and fats. Nonetheless, peaches contain significant amounts of fiber, ascorbic acid, and vitamin E. Consuming peaches can have positive effects on health due to their diverse array of medicinally significant substances, including carotenoids, polyphenolics, and antioxidants. Peach consumption has also been associated with various therapeutic benefits, such as anti-diabetic, anti-cardiovascular, and chemo preventive effects [39] (Figures 2.29 and 2.30).

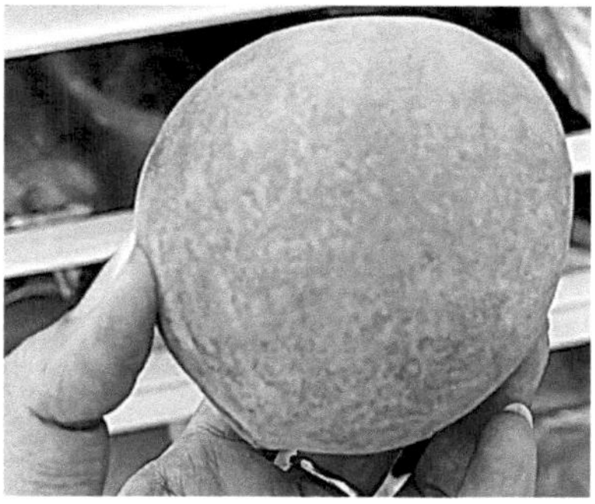

**FIGURE 2.29** Peach fruit.

**FIGURE 2.30** Peach fruit on tree.

*Source*: Photo by Jordan on Unsplash.

### 2.5.14.1 NUTRITIONAL PROFILE

Since the fruit is 87% water, it has a comparatively low caloric content. However, it is rich in carotenoids, particularly beta-carotene, alpha-carotene, and cryptoxanthin, which are precursors to vitamin A. In stone fruit, the main phenolic constituents are chlorogenic acid, epicatechin, and catechin. Carotenoids are said to be the most pervasive pigments in nature and are the reason why everything from yellow to red is the color it is. These fruits have a lot of lutein, zeaxanthin, beta-carotene, and alpha-carotene as their main carotenoids [39].

Peaches are a reasonably good source of carbohydrates compared to drupes. Fruit contains 9.54 g of total carbohydrates per 100 g of fruit weight. The majority of these carbs are found in fruit as sugars, and a peach typically has a sugar level of roughly 7 g/100 g. Fruit primarily contains sucrose, followed by glucose and fructose. Peaches additionally provide sorbitol, xylose, maltose, and galactin addition to these. Peaches are virtually cholesterol- and triglyceride-free. The majority of the fat in peaches is mono- and polyunsaturated. Oleic and linoleic fatty acids make up the majority of peaches' fat content [39].

Peaches are regarded as an excellent source of fiber. They have a fiber level that ranges from 1.4 to 2.4%. Depending on the cultivar, peaches also have a high pectin level that ranges from 1 to 1.5%. In addition to pectin, the fiber content of peaches is composed of hemicellulose, cellulose, xylose, and lignan. Peaches acquire their distinctive flavor from organic acids. Malic, citric, fumaric, and tartaric acids are among the organic acids that contribute to its flavor. Additionally, phenolic acids including p-coumaric, procyanidins b, neochlorogenic acid, catechinic acid, vanillic acid, chlorogenic acid, epicatechin, rutin, quercetin, and kaempferol. Malic and citric acid, however, are the two main organic acids in peaches that are found in equal amounts. Peaches contain a very small quantity of protein (1.2%) and none of the necessary amino acids. The major amino acid in peaches, alanine, is what gives them their distinctive flavor. Peach protein meets 3% of the total daily requirement for protein [39].

The color and flavor of fruit are influenced by polyphenolics, which are widely dispersed throughout plant tissue. Peaches are a significant source of polyphenols, with the main ones being chlorogenic acid, epicatechin, and catechin. In addition to these, flavanols, gallic acid, ellagic acid, etc., are also present. The fruit's color is attributed to carotenoids, a group of antioxidants. The main carotenoids found in peaches include lutein, beta-carotene, zeaxanthin, and cryptoxanthin. Peaches are a great source of water-soluble B complex vitamins, including vitamins A, C, and E. Peaches have a higher percentage of vitamin A (7%) compared to pears. Peaches also contain a significant amount of potassium, similar to other fruits. Additionally, peaches provide good amounts of magnesium, phosphorus, zinc, copper, manganese, iron, and calcium, while having low levels of sodium [39].

### 2.5.15 STRAWBERRY

Strawberries, which belong to the *Fragaria* genus in the *Rosacea* family, are well-known for their delicious flavor and exceptional red color. While they are low in protein and fat, they are high in important vitamins and minerals. Strawberries are particularly abundant in vitamin C, accounting for a large chunk of the necessary daily consumption in only one serving. They also include dietary fiber, folate, potassium, and manganese. Furthermore, strawberries are high in antioxidants such as flavonoids, phenolic acids, and anthocyanin, which add to their health benefits. Regular eating of strawberries has been linked to a variety of health advantages, including enhanced heart health owing to its capacity to decrease blood pressure and cholesterol. Furthermore, its antioxidant concentration may help decrease inflammation and oxidative stress in the body, thereby decreasing the risk of chronic illnesses like as cancer. Furthermore, strawberries contain ellagic acid, which has anti-cancer effects. Overall, strawberries are not only tasty, but they also provide several nutritional and health benefits, making them an important complement to a well-balanced diet (Figures 2.31 and 2.32).

#### 2.5.15.1 NUTRITIONAL PROFILE

Strawberries have a low-calorie content and are free from fat, cholesterol, and sodium. They are also high in fiber, vitamins, and polyphenols. Strawberries are often a favorite fruit for dessert, such as in the classic American dessert, strawberry shortcake. Fresh strawberries are low in carbs due to their high water content. They contain 7.68 g of total carbohydrates per 100 g of fruit. The majority of these carbs come from simple sugars like fructose, glucose, and sucrose, with fructose being the most abundant. The total sugar content in strawberries is 4.89 g/100 g of fruit. Strawberries have a net digestible carbohydrate content of less than 6 g per 100 g and a low glycemic index of 40 [39].

**FIGURE 2.31** Strawberry fruit.

**FIGURE 2.32** Strawberry fruits on tree.

*Source*: Photo by Dovydas Žilinskas on Unsplash.

The protein content in strawberries is extremely low, with only 10.67 g/100 g. They also lack some important amino acids like methionine. Other fruits also have subpar protein content. Glutamic acid is a common amino acid found in strawberries. Strawberries contain about 2 g of fiber per 100 g of fruit, which accounts for 26% of the total carbohydrates. The daily value (DV) for fiber in strawberries is 10%. The ratio of soluble to insoluble fiber is 0.28. With a total fat level of 0.29 g, strawberries are considered a low-fat food, like other fruits. The small amount of fat present is mostly polyunsaturated. The main fatty acid in strawberries is linoleic acid (omega 6), with a DV of 4%. Strawberries have a total mineral concentration of 0.49 g/100 g. They are rich in potassium, phosphorus, magnesium, calcium, and chloride, as well as trace amounts of boron, iodine, and selenium. Manganese has the highest DV among the minerals present [24].

Strawberries are abundant in vitamin C, with an ascorbic acid concentration of approximately 84.67 mg per 100 g. They have the highest DV (113%) for any micronutrient or macronutrient. Strawberries also contain other water-soluble vitamins like choline, B12, and B9. However, they have relatively low levels of fat-soluble vitamins like vitamin K and E. Strawberries are a rich source of folate, an essential vitamin for good health and the treatment of illnesses. Citric acid is the main organic acid in strawberries, followed by malic and vitamin C. Organic acids help give the apple its distinctive flavor and tartness. Strawberries contain more than 25 different types of anthocyanidins. The primary anthocyanin in strawberries, pelargonidin, is what gives them their color. Strawberries include significant levels of ellagitannins and ellagic acid. The primary anthocyanins in strawberries are pelargonidin and procyanidins [39].

Strawberry phenolic concentration is 2–10 times greater than that of other fruits, making them one of the top sources of phenolics. In strawberries, ellagitannins the sanjin is the predominant ellagitannin found in strawberries, followed by flavanols (such as quercetin and kaempferol-3-malonyl glucoside), flavanols, and phenolic acids [39].

### *2.5.16 CITRUS FRUITS*

Citrus is the collective name for all fruits in the Rutaceae family, which has the largest genus of any fruit family. The citrus fruits in this genus that are most valuable commercially are lemon, lime, sweet and sour oranges, tangerines, grapefruit, citron, and shaddock. The sweet orange, across all species, is the largest fruit crop of this genus, accounting for around 70% of the world's total citrus production. The high nutritional profile of citrus fruit makes it one of the most popular flavors in the world. Citrus fruits are well-known for their abundance of vitamins, fiber, and pectin, all of which are necessary for the body's normal functioning. Additionally, citrus fruits contain other physiologically active non-nutrient molecules that support the body's various metabolic activities seamlessly [39] (Figure 2.33).

**FIGURE 2.33** Different citrus fruits.

*Source*: Photo by Monika Grabkowska on Unsplash.

### *2.5.16.1 NUTRITIONAL PROFILE*

The high nutritional profile of citrus fruit makes it one of the most popular flavors in the world. Citrus fruits have a significant amount of vitamins, followed by minerals, fiber, and pectin, all of which are necessary for the body's normal operation. In addition to these, numerous physiologically active non-nutrient molecules that support the body's many metabolic processes are also found in significant concentrations [39].

Carbohydrates: Citrus fruits all have a similar amount of carbohydrates, with tangerines having the most (13.3 g/wg) and lemons having the least (9.32). Together, the three simple sugars fructose, glucose, and sucrose make up about 80% of the carbohydrates in an orange. In all citrus fruits, the ratio of sucrose to glucose to fructose is typically 2:1:1. However, lowering sugars make up the majority of the sugars in lemons.

Protein: Citrus fruits often contain little protein. However, compared to other fruits in the genus, lemons have somewhat more protein. Oranges contain between 0.068 and 0.120 g of total nitrogen per 100 g. Citrus fruits have produced more than 47 distinct enzymes that have been identified. These fruits have also been found to contain other phenolic amines, such as "synephrine."

Amino Acids: Citrus fruit has a low profile of amino acids. Citrus fruits contain a large amount of non-essential amino acids. The only semi-essential amino acid that is present in quite significant quantities is arginine.

Fatty Acids: Omega-6 makes up 39% of the fruit's total fat content, making it the main fatty acid in citrus fruits. Additionally, there are significant levels of oleic (37%) and palmitic (25%) acids. There are also negligible levels of linoleic (18%) and palmitoleic (7%) acids.Fiber: Citrus fruits are fruits with a high fiber content [39].

Citrus fruit's fiber is primarily made up of substances that are insoluble in alcohol. Lemon has the highest amount of fiber (2.80 g per 100 g), while grapefruit has the lowest (1.60 g per 100 g) in the citrus genus. Pectin, which makes up roughly 65–70% of the total amount of fiber in citrus fruits, is the most prevalent type of fiber. When associated with pectin from apples (6–9%), the methyl concentration of pectin from citrus is higher (7–10%). By reducing the amount of time food needs to travel through the GI system, dietary fiber avoids digestive issues. Additionally, pectin is known to decrease cholesterol. In general, citrus fruit peels are a great source of dietary fiber [39].

They are crucial for fruit flavor. Citric acid is the most prevalent organic acid in citrus. This organic acid accounts for around 10% of the

TSS, or total acidity, in matured oranges, 90% in grapefruit, and 100% in lemons. Other citrus fruits, excluding lemon and lime, also include the acids malic, oxalic, malonic, and quinic. The mineral that is most prevalent is potassium. Citrus fruits, like a select few others, are low in salt and high in potassium. Citrus fruits also contain some of the divalent ions Ca and Mg, albeit in lesser levels. Citrus fruits also include the trace elements Cu, Zn, Fe, and Mn. Citrus fruits are rich in ascorbic acid, a necessary water-soluble vitamin, which is the most significant nutrient. Lemon and orange both have the greatest levels of vitamin C of any fruit. Tangerines have the lowest ascorbic acid content of any citrus fruits. However, the peel contains the majority of the vitamin C—75% in oranges and 56% in grapefruit. A grapefruit or orange has 56–70 mg of ascorbic acid in it. Additionally, when citrus fruit ages, its ascorbic acid level rises. However, apart from vitamin C, the only one vitamin found in significant amounts in citrus fruit is retinol [39].

Vitamin A is mostly found as a carotenoid. Fruit color is determined by both forms of carotene, as well as cryptoxanthins. Citrus fruits contain a significant amount of cryptoxanthin. Aside from these, citrus fruit provides a negligible amount of vitamin E, while vitamin D is completely absent. Citrus fruits contain a high concentration of phytonutrients such as limonoids, hydroxyl benzoic and hydroxyl cinnamomic acid, and acetophenones. More than 40 limonoids have been recognized in citrus fruit, the most common of which are limonin and nemaline. Grapefruit and orange have the highest concentrations of these two primary limonoids among citrus fruits. Citrus fruits also contain a significant amount of flavonoids. Citrus flavonoids include quercetin and myricetin. Flavonoids, which have a polyphenolic ring, contribute to the flavor of several vegetables and fruits, and possess fungicidal potential. The fruits of citrus contain flavanones, a well-known type of flavonoids. Along with limonoids and quercetin, naringin, a significant flavanone, is accountable for the unpleasant flavor of citrus fruits. Naringin hydrolysis by gut bacteria prior to absorption results in the formation of naringenin, a precursor to various flavonoids. Naringin has been linked to numerous health benefits, including antioxidant activity and a role in modifiable cytochrome P450 enzymes. According to recent studies, hesperidin may have potential cardiovascular disease preventive properties [39].

### 2.5.17 MANGO (MANGIFERA INDICA)

Mango is the most significant and attractive fruit economically, and it is properly referred to as the "King of Fruits." It is generally accepted due to its distinct foreign flavor, taste, and nutritional benefits. Mango is high in nutrients, including carbs, soluble fiber, vitamins, free radical scavenging containing compounds, and minerals. It is high in polyphenols and carotenoids, both of which have significant benefits for health. Its fruits are also famous for their free radical scavenging activity, as well as a variety of other health advantages, including anti-inflammatory, anti-carcinogenicity, hepatoprotective, and antibacterial action. Mangoes may be used as functional foods to promote health and well-being [39] (Figures 2.34 and 2.35).

#### 2.5.17.1 NUTRITIONAL PROFILE

Mango is high in nutrients, including polysaccharides, soluble fiber, vitamins, minerals, and free radical scavengers. It is also high in polyphenols and carotenoids. Mangoes contain citric, malic, and organic acids. The carbohydrate content of mango is 14 g/100 g. The main sugars found in mango fruit are glucose and fructose. The quality of mango is determined by the sugar production during maturation, which is influenced by the amount of starch present in the cultivar. Mango pulp has more starch than the peel, while the mango kernel contains 58% starch on a dry basis. Mango peel is rich in pectin (0.8%), which contains neutral sugar residues that inhibit gel formation. Mango pulp and byproducts contain both soluble and

**FIGURE 2.34** Mango fruit.

*Source*: Photo by Ivar Leidus. https://creativecommons.org/licenses/by-sa/4.0/

**FIGURE 2.35** Mango fruit on tree.

insoluble dietary fiber. The complex carbohydrate portion of the dietary fiber is mainly composed of glucose, arabinose, and galactose [39].

Tropical fruits have a low-fat content. Mango lipids range from 0.20 to 0.38 g per 100 g. Mango kernels are rich in saturated fats, particularly stearic and palmitic acids. The main unsaturated fatty acids found in mango kernel oil are oleic, linoleic, and LAs. Arachidic acid is present in 1–2% of mango cultivars. Mango is also a good source of micronutrients and minerals [39].

Phytochemical compounds Mango variants are high in phenolic compounds, which function as radical scavengers and have a diversity of pharmacological potential. Gallic acid and its derivatives are all found in mango phenolic compounds. Mango peel is especially high in phytochemicals. Unripe mangoes have a greater phenolic content than fully ripe mangoes. Mango's primary organic acids are malic acid and citric acid [39].

## 2.5.18 GUAVA (PSIDIUM GUAJAVA)

Guava, one of the most sociable fruits, is widely consumed across the globe. It is the berry of the myrtle family and the genus Psidium. Its fruit is known as the "Queen of Fruits" because it has a high concentration of various nutrients. It is a fruit with a high protein content, a high fiber content, and no cholesterol. Guava, like other similar fruits, contains a high concentration of minerals and vitamins. The fruit has four times more vitamin C than oranges and 6–7 times more than other citrus fruits, three times more protein, and four times more fiber than pineapples [39] (Figures 2.36 and 2.37).

### 2.5.18.1 NUTRITIONAL PROFILE

Guava's abundance of macro- and micro-nutrients accounts for its numerous health benefits, such as supporting ocular health and controlling thyroid production, among others. Guava fruit, known as the "Queen

**FIGURE 2.36** Guava fruit.

**FIGURE 2.37** Guava fruit on tree.

*Source*: Photo by Anisa Wulan Asri on Unsplash.

of Fruits," is highly nutritious, being rich in protein and fiber while low in cholesterol. It also contains a high concentration of vitamins and minerals, with four times more vitamin C than oranges, three times more protein, and four times more fiber than pineapples. Additionally, guava has twice as much lycopene as tomatoes and a higher potassium content than bananas [39].

With a total carbohydrate content of 14.3 g and a nutritional value of 5%, guava has the lowest glycemic index (GI) among berries, scoring 33. The fruit contains approximately 9 g of sugars, mainly fructose. Guava is also a good source of protein among fruits, with a total protein level of 2.5 g/100 g fresh fruit weight. The primary amino acid in guava is glutamic acid, followed by leucine. Furthermore, guava is high in dietary fiber, with a total dietary fiber content of 5.4 g and a nutritional value of 22%. The berry has a dietary fiber level of 7.1 g/100 g insoluble and 1.8 g/100 g soluble. A high ratio of insoluble to soluble dietary fiber is beneficial for human health. Although guava contains a significant amount of fat (1.6 g/100 g), its nutritional value is low (2%) [39].

The fruits are high in vitamins and minerals, with the exception of vitamin D. Fruit has just 377 mg of ascorbic acid per 100 g. The nutritional value of vitamin C, on the other hand, is considerable, in the order of 628%, making the fruit a good source of vitamin C. Aside from ascorbic acid, retinol is abundant (1030 IU) and has a nutritional value of 21%. Guava also contains vitamin E, folate, choline, and other B complex vitamins. The ash content is approximately 1.4 g of the fresh fruit weight (100 g). Guava, like many fruits, is high in potassium and low in salt. Furthermore, guava is high in Fe, P, calcium, copper, and Mg. Guava also contains selenium, Zn, and other trace elements. Phytochemicals Guavas are high in polyphenols and carotenoids. Guava has a polyphenolic level of 2.62%, which translates to 126.40 mg/100 g of fresh weight of fruit. Condensed tannins make up the majority of the polyphenols (65%). Guava fruit contains more than 150 distinct beneficial components. Terpene hydrocarbons and 3-hydroxy 2 butanone contribute to the fruit's aroma. Guava terpenoids include galvanic acid and guava coumaric acid. The fruit's main component is carotene and lycopene. Furthermore, chlorophyll is present, although its concentration diminishes as ripening progresses. The most common are citric acid and malic acid, which are responsible for the acidity in fruit. Immature fruit has more citric acid, but malic acid comes first in later stages. Aside from these, the fruit contains lactic acid, galacturonic acid, coumaric acid, and gallic acid [39].

### *2.5.19 WALNUT (JUGLANS REGIA)*

Walnut is a digestible drupaceous nut known globally for its nutritional and health-promoting characteristics. Walnut kernels are abundant in nutrients and should be included in a well-balanced, health-promoting diet. It has a distinct flavor and an astringent taste. The nuts are rich in fat, protein, vitamins, and minerals, and they are an excellent source of flavonoids, phenolic acids, and related polyphenols. The kernel of the nut has a high concentration of antioxidants. Walnut kernel oil contains essential fatty acids, all of which have high antioxidant effects. Other documented medicinal benefits of walnuts include cardiovascular protection, improved cognitive function, and anti-inflammatory properties. In Asia and Europe, walnuts have traditionally been used to treat coughs, stomach problems, and cancer [39] (Figures 2.38 and 2.39).

**FIGURE 2.38** Walnut fruit.

**FIGURE 2.39** Walnut fruit on tree.

### 2.5.19.1 NUTRITIONAL PROFILE

Due to their high mineral, protein, vitamin, and fat content, walnuts are a nutrient-dense meal. Walnut kernels are a great source of polyphenols, phenolic acids, and related flavonoids. The digestible component of the walnut, the kernel, is encased in a brown seed coat and is a good source of free radical scavenging that shields the oil-rich seed from oxidation in the atmosphere. Walnut kernel oil includes important fatty acids that are widely utilized in the cosmetic sector due to their moisturizing and antioxidant characteristics. The sugar and carbohydrate content of walnuts is 13.7 g per 100 g and 2.6 g per 100 g of fruit weight, respectively. Walnuts' primary sugars include maltose, sucrose, and fructose. Walnuts have a low GI of 0 to 20 [39].

Due to the increased fat moieties of roughly 65% by weight in edible kernels, walnuts are categorized as calorie-rich meals. Triglycerides, free fatty acids, and sterol esters are the main constituents of walnut oil. Walnut oil contains the highest proportion of polyunsaturated fatty acids (PUFA), accounting for up to 80% of the total fat content. Palmitic and stearic acid are the main saturated fatty acids found in walnut kernels. LA, found in walnuts, is a necessary precursor for EPA and DNA 3 fatty acids, both of which have significant health advantages. Depending on the cultivar, the protein level of walnuts ranges from 14 to 25%. Walnuts, which have an amino acid score of 55, provide high-quality proteins [39].

Walnuts include insoluble dietary fiber, which aids in weight maintenance. They contain 2% protein by dry weight (DW) and contribute to dietary fiber. Walnuts are rich in minerals and vitamins that promote health, such as vitamin E. Tocopherol is one of the components of walnuts. They have the highest antioxidant activity among all foods. In addition to tocopherol, walnut kernels contain retinol, vitamin B6, niacin, pyridoxine, and vitamin K. Vitamin B6 promotes the health of the immune and neurological systems. The most abundant mineral in walnuts is vitamin K, followed by sodium, phosphorus, magnesium, calcium, zinc, iron, and copper. Walnuts contain 21 mg of gamma tocopherol per 100 g, which is 140% of the recommended daily intake. The skin of walnuts has a high concentration of phytochemicals and antioxidants. Ellagic acid and its derivatives, such as ellagitannins, reduce the risk of heart disease. Flavonoids have strong free radical scavenging properties. Walnuts also contain melatonin, a neurohormone that promotes sleep and reduces free radical damage. The most common phytonutrients found in walnuts are phenolic acid, flavonoid, hexahydroxydiphenic acid, and gallic acid, which are used to esterify hydrolysable tannins. Strictinin, a tannin, is also found in walnuts [39].

### 2.5.20 ALMOND (*PRUNUS DULCIS*)

Although classified as a nut, almonds are botanically classified as drupes of the Rosaceae family and sub-genus Amygdalus. The fruit is a key crop in the state of Jammu and Kashmir. Because of their high nutritional profile, almonds are ranked as the best and most nutritious food by the FDA. They are low in cholesterol and saturated fatty acids, but high in vital amino acids, omega-3 fatty acids, and phenolics. Almonds are also classified as a high-fiber food under European regulations. Amygdalin, a bioactive molecule found in almonds, is renowned for its possible health benefits such as cholesterol reduction and body weight regulation. Furthermore, due to their high fatty acid content, almonds are advantageous to those suffering from hypercholesterolemia. Additionally, the occurrence of biologically active phytochemicals in almonds has been shown by the American Heart Association to have beneficial effects on individuals who are at risk of developing CVD. Almonds are also beneficial against heart-related diseases, type 2 diabetes, and cancers [39] (Figures 2.40 and 2.41).

**FIGURE 2.40** Almond fruit.

**FIGURE 2.41** Almond fruit on tree.

*Source*: Photo by Plàcid Pérez Bru. https://creativecommons.org/licenses/by-sa/3.0/

### *2.5.20.1 NUTRITIONAL PROFILE*

Because of its high nutritional value, almonds have been designated as a nutrient-dense food by the FDA. Almonds are a cholesterol-free nut and are rich in nutrients. A 100 g serving of almonds provides 575 calories, 12.2 g of fiber, and a significant amount of alpha-tocopherol, total fat, protein, potassium, magnesium, phosphorus, calcium, and iron. They have a low amount of saturated fat but are high in other antioxidants. Almonds have a total carbohydrate content of 21.6 g per 100 g, accounting for 22% of the weight of the fresh fruit. The sugar content of almonds is approximately 4.4 g, consisting of sucrose, maltose, glucose, fructose, and galactose. Almonds are classified as low-sugar foods according to European regulations, as their sugar level is less than 5 g per 100 g. Additionally, almonds contain 0.7 g of starch in addition to sugars [39].

Among nuts, almonds are a rich source of high-quality protein, containing all necessary amino acids. The protein concentration in almonds is approximately 21.2 g per 100 g, which is equivalent to 21% of the weight of the kernel. Each almond seed contains 0.25 g of protein. Amandine is the primary storage protein in almonds and is considered an allergen, accounting for more than 70% of the total soluble protein. A 100 g serving of almonds provides 12.5 g of fiber, and almonds are classified as a high-fiber food according to European regulations. A serving of 28–30 g of almonds fulfills 14% of the daily fiber requirement [39].

Almonds are a good source of lipids. Almonds have a total fat proportion of 50 g per 100 g. Despite being rich in lipids, consuming almonds is considered healthy due to the higher proportion of monounsaturated fatty acids (31.6 g), followed by PUFA (12.3 g), and the lowest saturated fat content (3.7 g per 100 g) compared to other nuts. Additionally, almond kernels consist of 62% monounsaturated fatty acids and 24% PUFA. The primary monounsaturated fatty acid found in almonds is oleic acid (30.61 mg/100 g), while the major polyunsaturated fatty acid is linoleic acid (12.06 mg per 100 g). Almonds are also high in alpha-tocopherol (25.6 mg per 100 g) and choline (52.1 mg per 100 g). In addition to these, almonds provide vitamin B complex, but lack ascorbic acid, vitamin D, and K. Almonds have an ash level of about 3 g per 100 g. They are rich in calcium (264 mg per 100 g) and magnesium (268 mg per 100 g). Almonds are considered high-potassium, low-sodium meals with traces of selenium (2.50 g). Almonds contain various phytonutrients, including phenolic acids, phytosterols, and polyphenolic substances. Phenolic acids and tannins are among the polyphenols found in tree nuts. Furthermore, almonds, like other tree nuts, are low in dietary cholesterol but high in phytosterols. Phytosterols are substances that help maintain stable blood cholesterol levels by limiting cholesterol absorption. The most prevalent phytosterols in almonds are campesterol, sitosterol, 5-avenasterol, and stigmasterol [39].

### 2.5.21 PECAN NUT (CARYA ILLINOENIS)

Pecan is the fruit of a deciduous tree in the Juglandaceae family of the genus Carya. Pecans, like other fruits in the Juglandaceae family, are technically drupes rather than true nuts. Due to lipids and carbs accounting for 93% of pecan kernel weight, it is considered a high-energy dietary source. Additionally, the fruit is rich in protein, iron, dietary fiber, manganese, magnesium, phosphorus, zinc, and thiamine. Its low sodium and high potassium composition make it recommended for individuals with high blood pressure. Moreover, it aids in weight management by reducing blood lipids, atherosclerosis, and blood sugar levels [39] (Figures 2.42 and 2.43).

**FIGURE 2.42** Pecan nut fruit.

*Source*: Photo by Sara Cervera on Unsplash.

**FIGURE 2.43** Pecan tree in fruiting season.

### 2.5.21.1 NUTRITIONAL PROFILE

Value Pecans are a high-energy food item. Typically, lipids and carbohydrates account for around 93% of pecan kernel weight. Pecans contain 691 calories per 100 g and more than 100% of the DV for total fat. Pecans have the highest fat content among all nuts. Pecans are rich in fiber, manganese, magnesium, phosphorus, zinc, and thiamine. They are also high in protein, iron, and B vitamins. The majority of their fat composition consists of monounsaturated fatty acids, particularly oleic acid, and polyunsaturated fatty acids, primarily linoleic acid. Additionally, pecan nuts have the highest overall concentration of polyphenols among all nuts. Pecans have a lower protein content compared to other tree nuts such as almonds, walnuts, cashews, hazelnuts, and pistachios. Depending on the variety, pecan kernels typically contain 9–18% protein. The amino acid score of pecans is 61 [39].

### 2.5.22 MULBERRY (*M. RUBRA*)

Mulberry, also known as Sangzhi or Ramulus Mori, is a native plant of China. In terms of acreage, India has the second largest cultivation of mulberry. The fruit has been used as a folk medicine for hundreds of years to treat sore throats, anemia, and tonsillitis due to its wonderful flavor, appealing color, low calorie content, and high nutritional value. Mulberry contains three main bioactive components – hydroxy stilbenes, resveratrol, and oxyresveratrol – which have neuroprotective and cardioprotective properties. Additionally, oxyresveratrol is known to inhibit tyrosinase and prevent the formation of melanin, a pigment responsible for skin hyper-pigmentation in patients [39] (Figures 2.44 and 2.45).

**FIGURE 2.44** Mulberry fruit.

*Source*: Photo by engin akyurt on Unsplash

**FIGURE 2.45** Mulberry fruit on tree.

*Source*: Photo by Dmitry Bukhantsov on Unsplash.

### 2.5.22.1 NUTRITIONAL PROFILE

Fruit has a carbohydrate value of 9.8 g per 100 g fresh weight and a total sugar level of roughly 8 g per 100 g fresh fruit weight. Berries' main sugars are glucose and fructose. The fruit is low in protein, with around 1.4 g of protein per 100 g of fruit. However, the amino acid composition of the fruit is good. Mulberry fruit contains 18 amino acids. This berry fruit has a 42% essential amino acid (EAA) to total amino acid (TAA) ratio, which is comparable to that of several high-quality protein sources like milk and fish. Lipids are found in small amounts in all fruits. Fruit contains 0.4 g of fat per 100 g of fresh fruit. However, this small amount of fat is polyunsaturated. Linoleic acid (18:2) is the dominant fatty acid in all mulberry genotypes, followed by palmitic acid (16:0). Mulberry has an acidic and sour flavor. Fruit has a pH of roughly 3.5, which contributes to its delicious flavor. The sour flavor of the fruit is caused by the acids present in it. Malic acid (123–218 mg/g) is the most abundant organic acid in mulberry, followed by citric acid (21–41 mg/g). The berry fruit contains a significant amount of ascorbic acid. Furthermore, it includes appropriate levels of vitamin A and several B complex vitamins. Mulberry also contains tocopherols and their precursors. Like many fruits, mulberry has a high potassium content. It is also rich in calcium, iron, and zinc. Mulberry leaves and fruits contain a high concentration of bioactive compounds such as alkaloids, anthocyanins, and flavonoids. Mulberry contains hydroxystilbenes, Resveratrol (trans-3,4,5-trihydroxystilbene), and Oxyresveratrol (trans-2,3,4,5-tetrahydroxystilbene). These bioactive compounds have strong neuroprotective and cardioprotective properties. Anthocyanins are primarily responsible for the color of mulberry fruits and have health-promoting properties.

### *2.5.23 SEA-BUCKTHORN (HIPPOPHAE TIBETANA)*

Sea-buckthorn is a deciduous fragrant shrub that grows in various environments, including sea beaches, mountains, dry deserts, and semiarid plateaus. It occurs naturally in the Ladakh region and is considered a significant plant for the sustainable economic growth of the local residents. The berries of the shrub are rich in vitamins, minerals, bioactive carbohydrates, and proteins. The seed oil contains high levels of palmitoleic acid, steroids, vitamin E, vitamin K, and carotenoids. Linoleic acid, alpha-linolenic acid, sitosterol, and palmitoleic acid are known for their immunological modulatory and apoptogenic properties. Sea-buckthorn seed oil also provides protection against hypoxia by inhibiting free radical production and reducing blood levels of malondialdehyde and catecholamines. Fruit extracts have shown to be effective in protecting against burns, ulcers, lesions, cutaneous, and mucosal injuries. Therefore, sea-buckthorn can be utilized as a valuable medicinal plant in addition to its uses as food, fuel, fodder, and for soil conservation [39] (Figure 2.46).

**FIGURE 2.46** Sea-buckthorn fruit on tree.

*Source*: Photo by Philipp Deus on Unsplash.

### 2.5.23.1 NUTRITIONAL PROFILE

Sea-buckthorn berries are rich in vitamins, minerals, bioactive compounds, carbohydrates, and proteins. The seed oil is abundant in palmitoleic acid, steroids, vitamin E, vitamin K, and carotenoids. The fruit and its byproducts are also used for oil extraction. The oil is high in oleic acid, linoleic acid, LA, omega 3, omega 6, and omega 9 fatty acids. Sea-buckthorn leaves, with a protein content of 20%, are utilized as animal feed. The leaves also contain sterols, carotenoids, and triterpenes, which have medicinal properties. The main fruit acids responsible for astringency are quinic acid and malic acid. The flavor of sea-buckthorn berries varies depending on the cultivar, agroclimatic conditions, and harvesting period, with sugars giving them a pleasing taste. The berries contain 90% of their total sugar content in the form of glucose and fructose. Additionally, there are trace amounts of sugar alcohols such as sorbitol and mannitol [39].

The oil produced from sea-buckthorn seeds contains 80% to 85% unsaturated fatty acids, with 40–45% polyunsaturated fixed oil and 35–45% monounsaturated fatty acids. It is rich in LA, palmitoleic acid, oleic acid, and alpha-linolenic acid. The dried leaves of the sea-buckthorn plant contain 21.1 g of protein per 100 g, while the dried berries contain 20 g of protein per 100 g. Additionally, the berries contain nearly 19 different types of amino acids, 8 of which are essential amino acids. Common sea-buckthorn is also abundant in minerals and vitamins, including iron, calcium, magnesium, manganese, hydrogen, sodium, phosphorus, ascorbic acid, vitamin B2 riboflavin, and zinc [39].

Malic acid and quinic acid are the most abundant organic acids in sea-buckthorn, accounting for 90% of the total fruit acid. The acidity of sea-buckthorn berries is inversely related to their sweetness. Malic acid contributes significantly to the sourness and astringency of sea-buckthorn berries and juice. The higher the percentage of malic acid in the fruit, the more sour and astringent it becomes. Sea-buckthorn berries and juice are also rich in phenolic compounds. The berries are high in phytosterols, particularly sitosterol and campesterol, which help lower blood cholesterol levels. Sea-buckthorn contains approximately 300 phytochemicals of various classifications, including carotenoids, quinines, tocopherols, and other phytochemical constituents [39].

## 2.6 SUMMARY

Himalayan fruits contain antidepressant chemicals, which provide a natural way to fight depression. The special soil and temperature of the Himalayas allow these fruits to grow to their full potential, and their bioactive components have been associated with positive psychological impacts. Himalayan fruits are rich in antioxidants, vitamins, minerals, and other beneficial substances that may aid enhance mental health and reduce symptoms of depression. Evidence suggests that vitamins, polyphenols, and flavonoids all play a part in mood regulation via their ability to modify neurotransmitter levels, decrease inflammation, and protect against oxidative stress. A natural and nourishing approach to supporting mental health and fighting depression might be to include Himalayan fruits in one's diet.

## KEYWORDS

- **antidepression agents**
- **brain**
- **depression**
- **fruits**
- **Himalayan**
- **neurons**
- **nutrients**
- **phytochemicals**
- **vitamins**

## REFERENCES

1. Abadía, J., López-Millán, A. F., Rombolà, A., & Abadía, A., (2002). Organic acids and Fe deficiency: A review. *Plant and Soil, 241*, 75–86.
2. Ahmed, A., (2011). Apple phytochemicals for human benefits. *IJPR, 1*, 40.
3. Alnuwaiser, M. A., (2017). Study and comparison of amino acids in red and green apples. *Int. J. Pharm. Sci. Res., 8*, 4097–4105.

4. Alpert, J. E., & Fava, M., (1997). Nutrition and depression: The role of folate. *Nutrition Reviews, 55*, 145–149.
5. Aslam, H., Green, J., Jacka, F. N., Collier, F., Berk, M., Pasco, J., & Dawson, S. L., (2020). Fermented foods, the gut and mental health: A mechanistic overview with implications for depression and anxiety. *Nutritional Neuroscience, 23*, 659–671.
6. Bak, L. K., Schousboe, A., & Waagepetersen, H. S., (2006). The glutamate/GABA-glutamine cycle: Aspects of transport, neurotransmitter homeostasis and ammonia transfer. *Journal of Neurochemistry, 98*, 641–653.
7. Bauernfeind, J., (1972). Carotenoid vitamin A precursors and analogs in foods and feeds. *Journal of Agricultural and Food Chemistry, 20*, 456–473.
8. Bell, A. C., & D'Zurilla, T. J., (2009). Problem-solving therapy for depression: A meta-analysis. *Clinical Psychology Review, 29*, 348–353.
9. Bell, I. R., Edman, J. S., Morrow, F. D., Marby, D. W., Mirages, S., Perrone, G., Kayne, H. L., & Cole, J. O., (1991). B complex vitamin patterns in geriatric and young adult inpatients with major depression. *Journal of the American Geriatrics Society, 39*, 252–257.
10. Bhatt, S., Nagappa, A. N., & Patil, C. R., (2020). Role of oxidative stress in depression. *Drug Discovery Today, 25*, 1270–1276.
11. Blanco-Gomis, D., Mangas, A. J. J., Cabrales, I. M., & Abrodo, P. A., (2001). Gas chromatographic analysis of total fatty acids in cider. *Journal of Agricultural and Food Chemistry, 49,* 1260–1263.
12. Bourne, M., (1968). Texture profile of ripening pears. *Journal of Food Science, 33,* 223–226.
13. Boyer, J., & Liu, R. H., (2004). Apple phytochemicals and their health benefits. *Nutrition Journal, 3,* 1–15.
14. Brendgen, M., & Poulin, F., (2018). Continued bullying victimization from childhood to young adulthood: A longitudinal study of mediating and protective factors. *Journal of Abnormal Child Psychology, 46*, 27–39.
15. Brown, A., & Gershon, S., (1993). Dopamine and depression. *Journal of Neural Transmission/General Section JNT, 91*, 75–109.
16. Castillo, S. O., Baffi, J. S., Palkovits, M., Goldstein, D. S., Kopin, I. J., Witta, J., Magnuson, M. A., & Nikodem, V. M., (1998). Dopamine biosynthesis is selectively abolished in substantia nigra/ventral tegmental area but not in hypothalamic neurons in mice with targeted disruption of the Nurr1 gene. *Molecular and Cellular Neuroscience, 11*, 36–46.
17. Challem, J., (2011). *AARP The Food-Mood Solution: All-Natural Ways to Banish Anxiety, Depression, Anger, Stress, Overeating, and Alcohol and Drug Problems – and Feel-Good A*. John Wiley & Sons.
18. Chen, J., Wang, Z., Wu, J., Wang, Q., & Hu, X., (2007). Chemical compositional characterization of eight pear cultivars grown in China. *Food Chemistry, 104*, 268–275.
19. Dantzer, R., (2016). Role of the kynurenine metabolism pathway in inflammation-induced depression: Preclinical approaches. *Inflammation-associated Depression: Evidence, Mechanisms and Implications,* 117–138.

20. De Escalada, P. M. F., Uribe, M., Fissore, E. N., Gerschenson, L. N., & Rojas, A. M., (2010). Influence of the isolation procedure on the characteristics of fiber-rich products obtained from quince wastes. *Journal of Food Engineering, 96*, 239–248.
21. De, L., & De, T., (2019). Healthy food for healthy life. *J. Glob. Biosci., 8*, 6453–6468.
22. Dias, G. P., Cavegn, N., Nix, A., Do Nascimento, B. M. C., Stangl, D., Zainuddin, M. S. A., Nardi, A. E., et al., (2012). The role of dietary polyphenols on adult hippocampal neurogenesis: Molecular mechanisms and behavioral effects on depression and anxiety. *Oxidative Medicine and Cellular Longevity, 2012.*
23. Eby, G. A., & Eby, K. L., (2006). Rapid recovery from major depression using magnesium treatment. *Medical Hypotheses, 67*, 362–370.
24. Ekong, M. B., & Iniodu, C. F., (2021). Nutritional therapy can reduce the burden of depression management in low-income countries: A review. *IBRO Neuroscience Reports, 11*, 15–28.
25. Folkerts, H., Michael, N., Tölle, R., Schonauer, K., Mücke, S., & Schulze-Mönking, H., (1997). Electroconvulsive therapy vs. paroxetine in treatment-resistant depression—A randomized study. *Acta Psychiatrica Scandinavica, 96*, 334–342.
26. Ford, A., Almeida, O., Flicker, L., Hirani, V., McCaul, K., Singh, U., & Van, B. F., (2015). B-vitamins and depression. *European Psychiatry, 30*, 1–1.
27. Franceschi, S., Dal Maso, L., Augustin, L., Negri, E., Parpinel, M., Boyle, P., Jenkins, D., & LaVecchia, C., (2001). Dietary glycemic load and colorectal cancer risk. *Annals of Oncology, 12*, 173–178.
28. Freeman, M. P., Fava, M., Lake, J., Trivedi, M. H., Wisner, K. L., & Mischoulon, D., (2010). Complementary and alternative medicine in major depressive disorder: The American Psychiatric Association Task Force report. *The Journal of Clinical Psychiatry, 71*, 3537.
29. Ghaffari, D. M., Hedayati, A., Khorasani, E., Bayati, M., & Keshavarz, K., (2020). Selective serotonin reuptake inhibitors in major depression disorder treatment: An umbrella review on systematic reviews. *International Journal of Psychiatry in Clinical Practice, 24*, 357–370.
30. Godswill, A. G., Somtochukwu, I. V., Ikechukwu, A. O., & Kate, E. C., (2020). Health benefits of micronutrients (vitamins and minerals) and their associated deficiency diseases: A systematic review. *International Journal of Food Sciences, 3*, 1–32.
31. Gorinstein, S., Zachwieja, Z., Folta, M., Barton, H., Piotrowicz, J., Zemser, M., Weisz, M., et al., (2001). Comparative contents of dietary fiber, total phenolics, and minerals in persimmons and apples. *Journal of Agricultural and Food Chemistry, 49*, 952–957.
32. Grosso, G., Galvano, F., Marventano, S., Malaguarnera, M., Bucolo, C., Drago, F., & Caraci, F., (2014). Omega-3 fatty acids and depression: Scientific evidence and biological mechanisms. *Oxidative Medicine and Cellular Longevity, 2014.*
33. Haen, C., & Aronson, S., (2017). *Handbook of Child and Adolescent Group Therapy*. New York, NY: Routledge.
34. Hanna, S., Lachover, L., & Rajarethinam, R., (2009). Vitamin B12 deficiency and depression in the elderly: Review and case report. *Primary Care Companion to the Journal of Clinical Psychiatry, 11*, 269.

35. Hansen, S. N., Ipsen, D. H., Schou-Pedersen, A. M., Lykkesfeldt, J., & Tveden-Nyborg, P., (2018). Long term westernized diet leads to region-specific changes in brain signaling mechanisms. *Neuroscience Letters, 676*, 85–91.
36. Hasegawa, H., & Nakamura, K., (2010). Tryptophan hydroxylase and serotonin synthesis regulation. *Handbook of Behavioral Neuroscience* (pp. 183–202). Elsevier.
37. Heinonen, M. I., Ollilainen, V., Linkola, E. K., Varo, P. T., & Koivistoinen, P. E., (1989). Carotenoids in Finnish foods: Vegetables, fruits, and berries. *Journal of Agricultural and Food Chemistry, 37*, 655–659.
38. Hudina, M., & Stampar, F., (2005). The correlation of the pear (*Pyrus communis* L.) cv.' Williams' yield quality to the foliar nutrition and water regime. *Acta Agriculturae Slovenica, 85*, 179–185.
39. Hussain, S. Z., Naseer, B., Qadri, T., Fatima, T., & Bhat, T. A., (2021). *Fruits Grown in Highland Regions of the Himalayas.* Springer.
40. Hussain, S. Z., Naseer, B., Qadri, T., Fatima, T., & Bhat, T. A., (2021). Pear (*Pyrus communis*)—Morphology, taxonomy, composition, and health benefits. *Fruits Grown in Highland Regions of the Himalayas* (pp. 35–48). Springer.
41. Hussain, S. Z., Naseer, B., Qadri, T., Fatima, T., & Bhat, T. A., (2021). Apples (*Pyrus malus*)—Morphology, taxonomy, composition, and health benefits. *Fruits Grown in Highland Regions of the Himalayas* (pp. 17–34). Springer.
42. Hussain, S. Z., Naseer, B., Qadri, T., Fatima, T., & Bhat, T. A., (2021). Apricots (*Prunus armeniaca*)—Morphology, taxonomy, composition, and health benefits. *Fruits Grown in Highland Regions of the Himalayas* (pp. 91–102). Springer.
43. Hussain, S. Z., Naseer, B., Qadri, T., Fatima, T., & Bhat, T. A., (2021). Quince (*Cydonia oblonga*)—Morphology, taxonomy, composition, and health benefits. *Fruits Grown in Highland Regions of the Himalayas* (pp. 49–62). Springer.
44. Hussain, S. Z., Naseer, B., Qadri, T., Fatima, T., & Bhat, T. A., (2021). Cherries (*Prunus cerasus*)—Morphology, taxonomy, composition, and health benefits. *Fruits Grown in Highland Regions of the Himalayas* (pp. 63–75). Springer.
45. Hussain, S. Z., Naseer, B., Qadri, T., Fatima, T., & Bhat, T. A., (2021). Fig (*Ficus Carica*)—Morphology, Taxonomy, composition and health benefits. *Fruits Grown in Highland Regions of the Himalayas* (pp. 77–90). Springer.
46. Hyman, S. E., (2013). Psychiatric drug development: Diagnosing a crisis. In: *Cerebrum: The Dana Forum on Brain Science*. Dana Foundation.
47. Janick, J., (2007). Fruits of the bibles. *HortScience, 42*, 1072–1076.
48. Jha, G. N., Sarma, D., & Clureshi, T., (2010). Natural sources of carotenoids and the beauty of Coldwater fish. *Coldwater Fisheries Management*, 333–344.
49. Josimuddin, S. K., Kumar, M., & Rastogi, H., (2022). A review on nutritional and medicinal value of malus domestica with various activity. *Int. J. Health Sci., 6*, 7251–7265.
50. Kadakal, Ç., Poyrazoglu, E. S., Artik, N., & Nas, S., (2004). Effect of activated charcoal on water-soluble vitamin content of apple juice. *Journal of Food Quality, 27*, 171–180.
51. Kumari, T., & Deka, S. C., (2021). Potential health benefits of garden pea seeds and pods: A review. *Legume Science, 3*, e82.

52. Leung, B. M., & Kaplan, B. J., (2009). Perinatal depression: Prevalence, risks, and the nutrition link—A review of the literature. *Journal of the American Dietetic Association, 109*, 1566–1575.
53. Lewis, J. E., Tiozzo, E., Melillo, A. B., Leonard, S., Chen, L., Mendez, A., Woolger, J. M., & Konefal, J., (2013). The effect of methylated vitamin B complex on depressive and anxiety symptoms and quality of life in adults with depression. *International Scholarly Research Notices, 2013*.
54. Li, X., Li, X., Wang, T., & Gao, W., (2016). Nutritional composition of pear cultivars (*Pyrus* spp.). *Nutritional Composition of Fruit Cultivars* (pp. 573–608). Elsevier.
55. Lintas, C., (1992). Nutritional aspects of fruits and vegetables consumption. *Options Mediterraennes, 19*, 79–87.
56. Liu, L., Liu, C., Wang, Y., Wang, P., Li, Y., & Li, B., (2015). Herbal medicine for anxiety, depression, and insomnia. *Current Neuropharmacology, 13*, 481–493.
57. Ma, S., Neilson, A. P., Lahne, J., Peck, G. M., O'Keefe, S. F., & Stewart, A. C., (2018). Free amino acid composition of apple juices with potential for cider making as determined by UPLC-PDA. *Journal of the Institute of Brewing, 124*, 467–476.
58. Maes, M., Leonard, B., Myint, A., Kubera, M., & Verkerk, R., (2011). The new '5-HT'hypothesis of depression: Cell-mediated immune activation induces indoleamine 2, 3-dioxygenase, which leads to lower plasma tryptophan and an increased synthesis of detrimental tryptophan catabolites (TRYCATs), both of which contribute to the onset of depression. *Progress in Neuro-psychopharmacology and Biological Psychiatry, 35*, 702–721.
59. Mariani, E., Polidori, M., Cherubini, A., & Mecocci, P., (2005). Oxidative stress in brain aging, neurodegenerative and vascular diseases: An overview. *Journal of Chromatography B, 827*, 65–75.
60. Markus, C. R., Olivier, B., & De Haan, E. H., (2002). Whey protein rich in α-lactalbumin increases the ratio of plasma tryptophan to the sum of the other large neutral amino acids and improves cognitive performance in stress-vulnerable subjects. *The American Journal of Clinical Nutrition, 75*, 1051–1056.
61. McCune, L. M., Kubota, C., Stendell-Hollis, N. R., & Thomson, C. A., (2010). Cherries and health: A review. *Critical Reviews in Food Science and Nutrition, 51*, 1–12.
62. McKeever, A., Agius, M., & Mohr, P., (2017). A review of the epidemiology of major depressive disorder and of its consequences for society and the individual. *Psychiatria Danubina, 29*, 222–231.
63. Moosavi, F., Hosseini, R., Saso, L., & Firuzi, O., (2016). Modulation of neurotrophic signaling pathways by polyphenols. *Drug Design, Development and Therapy, 10*, 23.
64. Moritz, B., Schmitz, A. E., Rodrigues, A. L. S., Dafre, A. L., & Cunha, M. P., (2020). The role of vitamin C in stress-related disorders. *The Journal of Nutritional Biochemistry, 85*, 108459.
65. Mudgil, D., & Barak, S., (2013). Composition, properties, and health benefits of indigestible carbohydrate polymers as dietary fiber: A review. *International Journal of Biological Macromolecules, 61*, 1–6.
66. Ng, F., Berk, M., Dean, O., & Bush, A. I., (2008). Oxidative stress in psychiatric disorders: Evidence base and therapeutic implications. *The International Journal of Neuropsychopharmacology, 11*, 851–876.

67. Owen, A. J., Batterham, M., Probst, Y. C., Grenyer, B. F., & Tapsell, L. C., (2005). Low plasma vitamin E levels in major depression: Diet or disease? *European Journal of Clinical Nutrition, 59*, 304–306.
68. Paredes, O. Z., (2011). Obesity Public Enemy# 1 or Death. Xlibris Corporation.
69. Park, Y. H., Choi-Kwon, S., Park, K. A., Suh, M., & Jung, Y. S., (2017). Nutrient deficiencies and depression in older adults according to sex: A cross sectional study. *Nursing & Health Sciences, 19*, 88–94.
70. Parletta, N., Milte, C. M., & Meyer, B. J., (2013). Nutritional modulation of cognitive function and mental health. *The Journal of Nutritional Biochemistry, 24*, 725–743.
71. Popa, T., & Ladea, M., (2012). Nutrition and depression at the forefront of progress. *Journal of Medicine and Life, 5*, 414.
72. Predictable, S., Laurencic, G., & Malone, D., (2006). Side effects of antidepressants: An overview. *Cleveland Clin. J. Med., 73*, 351.
73. Ptak, L. R., (1994). *The Effect of Indirect Dopamine Agonists on the Uptake of (3-H)-Dopamine in Embryonic Mesencephalic Tegmental Neuron Cultures*. Rush University, College of Nursing.
74. Ramadan, M. F., & Mörsel, J. T., (2003). Oil cactus pear (*Opuntia Ficus-indica* L.). *Food Chemistry, 82*, 339–345.
75. Rao, T. S., Asha, M., Ramesh, B., & Rao, K. J., (2008). Understanding nutrition, depression, and mental illnesses. *Indian Journal of Psychiatry, 50*, 77.
76. Rivest, R., Falardeau, P., & Di Paolo, T., (1995). Brain dopamine transporter: Gender differences and effect of chronic haloperidol. *Brain Research, 692*, 269–272.
77. Ross, C. E., Reynolds, J. R., & Geis, K. J., (2000). The contingent meaning of neighborhood stability for residents' psychological well-being. *American Sociological Review*, 581–597.
78. Sajid, S. M., Zubair, M., Waqas, M., Nawaz, M., & Ahmad, Z., (2015). A review on quince (*Cydonia oblonga*): A useful medicinal plant. *Global Veterinaria, 14*, 517–524.
79. Sánchez-Rodríguez, L., Ali, N. S., Cano-Lamadrid, M., Noguera-Artiaga, L., Lipan, L., Carbonell-Barrachina, Á. A., & Sendra, E., (2019). Flavors and aromas. *Postharvest Physiology and Biochemistry of Fruits and Vegetables* (pp. 385–404). Elsevier.
80. Sarris, J., Panossian, A., Schweitzer, I., Stough, C., & Scholey, A., (2011). Herbal medicine for depression, anxiety, and insomnia: A review of psychopharmacology and clinical evidence. *European Neuropsychopharmacology, 21*, 841–860.
81. Sharma, I. P., Kanta, C., Semwal, S., & Goswami, N., (2017). Wild fruits of Uttarakhand (India): Ethnobotanical and medicinal uses. *International Journal of Complementary & Alternative Medicine, 8*, 1–8.
82. Shoufeng, S., Juncai, L., Jun, W., & Shaoling, Z., (2011). Characteristics of organic acids in the fruit of different pear species. *African Journal of Agricultural Research, 6*, 2403–2410.
83. Silva, G., Souza, T. M., Barbieri, R. L., & Costa De, O. A., (2014). Origin, domestication, and dispersing of pear (*Pyrus* spp.). *Advances in Agriculture, 2014.*
84. Silver, F. W., & Ruckle, J. L., (1989). Depression: Management techniques in primary care. *Postgraduate Medicine, 85*, 359–366.
85. Singh, K., (2016). Nutrient and stress management. *J. Nutr. Food Sci., 6*, 528.

86. Slade, T., Johnston, A., Oakley, B. M. A., Andrews, G., & Whiteford, H., (2009). 2007 National Survey of Mental Health and Wellbeing: Methods and key findings. *Australian & New Zealand Journal of Psychiatry, 43*, 594–605.
87. Smeriglio, A., Alloisio, S., Raimondo, F. M., Denaro, M., Xiao, J., Cornara, L., & Trombetta, D., (2018). Essential oil of *Citrus lumia* Risso: Phytochemical profile, antioxidant properties and activity on the central nervous system. *Food and Chemical Toxicology, 119*, 407–416.
88. Smith, K., Fairburn, C., & Cowen, P., (1997). Relapse of depression after rapid depletion of tryptophan. *The Lancet, 349*, 915–919.
89. Soler, C., Soriano, J. M., & Mañes, J., (2009). Apple-products phytochemicals and processing: A review. *Natural Product Communications, 4*, 1934578X0900400504.
90. Sood, S., & Bhardwaj, M., (2015). Nutritional evaluation of quince fruit and its products. *Journal of Krishi Vigyan, 3*(2s), 67–69.
91. Southwick, S. M., Bremner, J. D., Rasmusson, A., Morgan, I. I. I., C. A., Arnsten, A., & Charney, D. S., (1999). Role of norepinephrine in the pathophysiology and treatment of posttraumatic stress disorder. *Biological Psychiatry, 46*, 1192–1204.
92. Stuart, S., (2012). Interpersonal psychotherapy for postpartum depression. *Clinical Psychology & Psychotherapy, 19*, 134–140.
93. Suganthy, N., Pandian, S. K., & Devi, K. P., (2009). Cholinesterase inhibitors from plants: Possible treatment strategy for neurological disorders: A review. *Int. J. Biomed. Pharm. Sci., 3*, 87–103.
94. Tag, H., Tsering, J., Hui, P. K., Gogoi, B. J., & Veer, V., (2014). Nutritional potential and traditional uses of high-altitude wild edible plants in Eastern Himalayas, India. *International Journal of Biological, Biomolecular, Agricultural, Food and Biotechnological Engineering, 8*, 238–243.
95. Trigueros, L., Pérez-Alvarez, J., Viuda-Martos, M., & Sendra, E., (2011). Production of low-fat yogurt with quince (*Cydonia oblonga* Mill.) scalding water. *LWT: Food Science and Technology, 44*, 1388–1395.
96. Tsepkova, P., Artiukhov, A., Boyko, A., Aleshin, V., Mkrtchyan, G., Zvyagintseva, M., Ryabov, S., et al., (2017). Thiamine induces long-term changes in amino acid profiles and activities of 2-oxoglutarate and 2-oxoadipate dehydrogenases in rat brain. *Biochemistry (Moscow), 82*, 723–736.
97. Wani, A. A., Singh, P., Gul, K., Wani, M. H., & Langowski, H., (2014). Sweet cherry (*Prunus avium*): Critical factors affecting the composition and shelf life. *Food Packaging and Shelf Life, 1*, 86–99.
98. Wojdyło, A., Oszmiański, J., & Bielicki, P., (2013). Polyphenolic composition, antioxidant activity, and polyphenol oxidase (PPO) activity of quince (*Cydonia oblonga* Miller) varieties. *Journal of Agricultural and Food Chemistry, 61*, 2762–2772.
99. Yu, J., & Ahmedna, M., (2013). Functional components of grape pomace: Their composition, biological properties, and potential applications. *International Journal of Food Science & Technology, 48*, 221–237.

## CHAPTER 3

# Role of Antidepressant Agents Derived from Selected Flowering Plants

KIRTI RAINA,[1] ARTI CHAUHAN,[1] ROHIT SHARMA,[2] SUNIL THAKUR,[3] ROHIT SHARMA,[4] and ASHUN CHAUDHARY[1]

[1]*Department of Plant Science, School of Life Sciences, Central University of Himachal Pradesh, Shahpur, Kangra, Himachal Pradesh, India*

[2]*Department of Forest Products, College of Forestry, Dr. Yashwant, Singh Parmar University of Horticulture and Forestry, Nauni, Solan, Himachal Pradesh, India*

[3]*Department of Animal Sciences, School of Life Sciences, Central University of Himachal Pradesh, TAB Shahpur, Kangra, Himachal Pradesh, India*

[4]*Department of Rasa Shastra and Bhaishajya Kalpana, Faculty of Ayurveda, Institute of Medical Sciences, Banaras Hindu University, Varanasi, Uttar Pradesh, India*

## ABSTRACT

Depression is a prevalent healthcare disorder worldwide. It not only affects a person's mood but can also disrupt sleep and eating patterns. It is more common in women, with twice the prevalence compared to men. In addition to impacting social behavior, depression can have detrimental effects on vital organs such as the brain, heart, and kidneys. Early signs of depression can vary among individuals and may include anxiety, memory loss, feelings

Himalayan Medicinal Plants for the Treatment of Depression: A Source of Rich Antidepressant Agents. Abdur Rauf, Prabhakar Semwal, & Hafiz Ansar Rasul Suleria (Eds.)

of sadness, weakened immune system, sudden changes in behavior, and in severe cases, suicidal thoughts may arise. The use of antidepressants has proven to be beneficial in treating anxiety and depression-related disorders. These drugs act on different neurotransmitter receptors and work by inhibiting the reuptake of neurotransmitters, thereby reducing the levels of serotonin, dopamine, and norepinephrine (NE). While there are various antidepressant drugs available in the market, they can have adverse side effects on the human body. Therefore, the use of plant-based antidepressant agents can be a promising alternative. Researchers have demonstrated the effective use of antidepressant agents derived from different parts of Himalayan plants in the treatment of depressive ailments. The main objective of this review is to compile descriptive information about the antidepressant agents found in flowering plants of the Himalayan region.

## 3.1 INTRODUCTION

Depression is a significant and serious psychological condition that affects approximately 17% of the global population. As directed by the World Health Organization (WHO), depression is projected to become the second leading cause of global illness burdens, impacting over 350 million people worldwide [10]. While grief and sadness are common human emotions that typically subside within a few days, acute depression persists for a longer duration [25]. Indications of depression can be categorized into two main components: biological and emotional. Biological signs include impaired thinking and behavior, loss of sexual desire, insomnia, and loss of appetite. Emotional symptoms encompass sadness, lethargy, lack of enthusiasm, low self-worth (including feelings of shame, incompetence, and ugliness), indecisiveness, and a sense of purposelessness. Depression is highly prevalent and causes more impairment than any other condition, making its prognosis and management challenging due to limited understanding of the disorder. Antidepressants are prescribed to counteract or alleviate depressive disorders and assist in managing various dependencies. While synthetic medications are commonly used as an effective therapy for individuals with depression, they can have adverse effects that may be hazardous. Prescribing synthetic drugs can be problematic if the patient also has other medical conditions, as the treatment of depression with synthetic antidepressants can be life-threatening. Due to over-reliance on antidepressant drugs, many individuals are using these treatments without informing doctors or seeking consent. The outcomes can be positive or negative, depending on the accumulation or manifestation of effects in patients with depressive disorders. Side effects of synthetic antidepressants include agitation, diarrhea, dizziness, dry mouth, excessive sweating, constipation, insomnia, increased heart rate, low blood pressure, and sexual problems such as loss of libido. In light of these circumstances, alternative treatments for depression utilizing therapeutic plants with antidepressant properties have gained attention. The treatment of depression and other mental difficulties with medication often leads to numerous side effects, prompting people to explore herbal plants with anti-depressive properties as a better alternative [74]. Depression can cause severe and life-threatening side effects, including suicidal tendencies, which are more common among adolescents. More than 50% of individuals experience their first symptoms after the age of 60. The onset of depressive symptoms is most frequent in early adulthood, between the ages of 20 and 40. Abrupt discontinuation of antidepressants can result in withdrawal symptoms such as nausea, diarrhea, dizziness, headache, irritability, sleep disturbances, nightmares, psychosis and seizures. In short-term investigations, in children and teenagers with depression and mental illnesses, antidepressants raised the likelihood of suicidal thinking and behavior. Hence these cannot be prescribed to children and teenagers with symptoms of depression. The Himalayan region has a diverse variety of plants with antidepressant properties. Since time immemorial, Scientists are continuously working on exploring the antidepressant properties found in the plants of the Himalayan region to cure depression.

According to the information provided by WHO, approximately 121 million people worldwide experience major depression [74]. By the year 2020, it is projected to become the second most common cause of death due to complications arising from stress, contributing to a rise in suicide cases by up to 15% [26]. Rough estimates indicate that 65.4 out of every 1,000 individuals in India have mental disorders. Mayo Clinic reports that more than one-third of Americans (13%) take antidepressants, with approximately one-fourth of women aged 50 to 64 using them. Antidepressants are the most widely prescribed class of medication, second only to antibiotics. The therapeutic ingredients found in a variety of plant extracts and their derived agents have shown antidepressant qualities [43]. Depression is caused by low levels of neurotransmitters such as serotonin, dopamine, and noradrenaline in the brain. Antidepressants work by influencing serotonin levels in the brain, blocking the reuptake of specific signaling chemicals and increasing their concentrations around nerves in the brain. The central and peripheral neural systems are connected to pain and depression, with the brain stem serving as a crucial link between the higher regions of the brain and the spinal cord. Serotonin and norepinephrine (NE), two neurotransmitters, play a role in pain perception. Ascending and descending brain pathways are involved in transmitting information. Depression affects approximately one out of every 20 individuals in a survey conducted by World Mental Health (WHM) in 17 countries over the past year. Given the numerous adverse effects of depression on mental health, it is crucial that we take this problem seriously. The wide range of available antidepressants makes it easier to select the appropriate medication based on an individual's symptoms and needs. For individuals who have experienced multiple episodes of major depression, long-term antidepressant treatment may be recommended to prevent further episodes. A family history of the disease is another factor that supports prolonged treatment. Antidepressant drugs can also help with anxiety, although their effectiveness varies from person to person. The duration of action of antidepressants in the body can vary, with some medications lasting approximately 36 hours and others lasting several days. While synthetic medications are commonly used to treat severe depression (Figure 3.1), they can have side effects that may compromise the therapeutic outcome, highlighting the potential for alternative plant-based natural drugs to be used as antidepressants.

India has a diverse range of medicinal plants found across the country in various geographical and ecological situations. Herbs have been utilized to cure a variety of ailments since ancient times. Before the advent of conventional medicine, medicinal herbs were used for centuries. The medicinal properties of these plants can be found in their phytochemicals, which have specific physiological effects on the human body. Plant-based pharmacological products have minimal or no side effects, making them the best alternative to synthetic anti-depressive medications. The discovery of natural products is an interesting area for identifying novel medications for psychological disorders, including depression and anxiety. Natural molecules have good biological accessibility and low toxicity. Phytochemical elements such as saponins, flavonoids, and alkaloids have been used as antidepressants. Currently, plants are used in the production of 30% of medicinal preparations. Different plant parts such as roots, stems, leaves, flowers, fruits, and seeds are used for medicinal purposes. It is

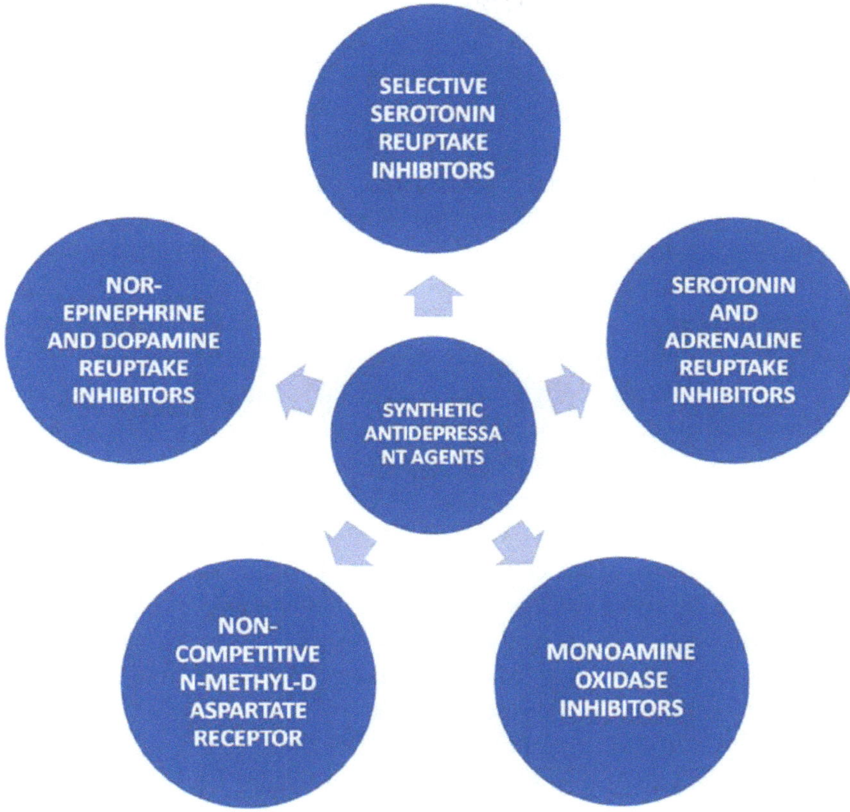


**FIGURE 3.1** Types of synthetic antidepressant agents.

interesting to note that the same plants can be used in various ways. The extract derived from the leaves of *Ocimum sanctum,* commonly referred to as Tulsi or Holy basil plant, exhibits calming and anti-aggressive effects on patients dealing with depression [93]. However, the antidepressive properties extracted from roots show more effectiveness compared to leaves. The leaf extract of the wild shrub, *Apocynum venetum Linn.*, contains major flavonoids such as hyperoside and isoquercetin, which play a significant role in depression therapy as antidepressant agents. Kaempferol, a natural flavonol (a type of flavonoid) present in the plant extract of *Crocus sativus,* commonly known as Saffron crocus, has shown improvement in cases related to depression [32].

## 3.2 REVIEW OF LITERATURE

A literature review is a scientific investigation of existing knowledge related to a specific area of interest. In this study, various antidepressant agents found in the Himalayan region were compiled. Singh et al. [88] reported that *Foeniculum vulgare* possesses various pharmacological activities and acts as a monoamine inhibitor by increasing NE, serotonin, and dopamine levels in the brain. Wasnik et al. [98] suggested that an alcoholic extract of *Phyllanthus amarus*, at a dosage of 50 and 100 mg/kg, exhibits antidepressant properties. Santosh et al. [80] demonstrated that a methanolic concentrate of passionflower leaves displayed behavioral responses in mice sensitive to therapeutically effective antidepressant drugs. Kumar et al. [48] found that the aerial part of *Abies pindrow,* containing flavonoids and terpenoids, exhibited anxiolytic properties suitable for treating depression. Rocha et al. [72] and Dhamija et al. [20] suggested that butanoic extracts (catechin and epicatechin) of *Cecropia glazioui* increased hippocampal monoamine levels while inhibiting serotonin, dopamine, and noradrenaline uptake by synaptosomes in several brain areas. The biochemical constituent kaempferol, found in *Crocus sativus*, was examined through a forced swimming test in mice, which revealed decreased immobility and verified its antidepressant properties. Ghasemi et al. [32] and Ma et al. [52] showed that *Rhodiola rosea* contains several biological compounds, such as flavonoids, glycosides, and phenolic acids. Majeed et al. [54] reported that *Hypericum perforatum* (Indian Hypericum) *Linn* has effects similar to those induced by imipramine and exhibits high antioxidant activity. Snitz et al. [90] suggested that the Chinese plant *Gingko biloba* has been used to treat depression without the adverse effects associated with synthetic antidepressants. Its combined effect with venlafaxine is effective at protecting neurons and minimizing brain damage. Hurley et al. [40] suggested that curcumin, present in *Curcuma longa*, possesses antidepressant properties in the WKY rat model, probably due to enhanced neurotrophic activity. Curcumin could be a long-acting and powerful natural antidepressant. *Perilla frutescensis* is frequently used to treat sadness and depression [4, 44]. In the forced swim and tail suspension test (TST), the length of immobility was dramatically reduced in a mouse model after oral intake of the plant essential oil. Sairam et al. [77] suggested that the methanolic concentrate of *Bacopa monniera* given orally to Charles-Foster albino rats (either sex) at oral management of 20 and 40 mg/kg for five days produced antidepressant-like effects in FST and models of learned helplessness, which is identical to imipramine at a concentration of 15 mg/kg. Lee et al. [50] demonstrated that "Piperine," a phytoconstituent extracted from an ethanol extract of *Piper longum* fruits, reduced immobility duration in TST substantially. It additionally suppressed Monoamine oxidase inhibitor-A and Monoamine oxidase inhibitor-B, indicating that it could be used as an antidepressant. Xu et al. [100] suggested that in male mice, a petroleum ether extract of Chinese plantain at a concentration of 5 and 10 mg/kg, showed considerable antidepressant-like effects. Herrera-Ruiz et al. [39] demonstrated that by reducing the period of immobility of mice and rats treated with FST, the leaves and flowers containing hydro-alcohol extract in *Salvia elegans* of the family *Lamiaceae* revealed antidepressant activity. Oshima et al. [66] informed that the methanolic root concentrate of *Valeriana fauriei* decreased the extent of male mice's immobility on the forced swimming test and had significant antidepressant properties. The active ingredients of the methanolic extract that were responsible for antidepressant activity were "Kessyl alcohol" and Guanine-type Sesquiterpenoids (Kessanol and Cyclokessyl acetate). The objectives of the review include a listing of Antidepressant flowers, identification of active compound from the literature survey, and different bioactivity associated with these antidepressant plants. The search strategy is illustrated in Figure 3.2.

## 3.3 MATERIAL AND METHODOLOGY

The research methodology includes the study of comprehensive literature available in online web search engines such as Google Scholar, Science Direct, PubMed, Dhara, and AYUSH, which were analyzed to obtain informative details about the list of various antidepressant agents found in various medicinal plants, along with their descriptions. The research is limited to achieving the listed objectives, which are to list antidepressant compounds in plants. The description of each plant includes the plant's binomial nomenclature, common name, family name, distribution, and the part of the plant that consists of various antidepressant agents. This study was conducted on 27/04/2022. English is preferred as the main language for conducting this study, as a significant amount of literature is available in English. The data processing can be done in the following manner.

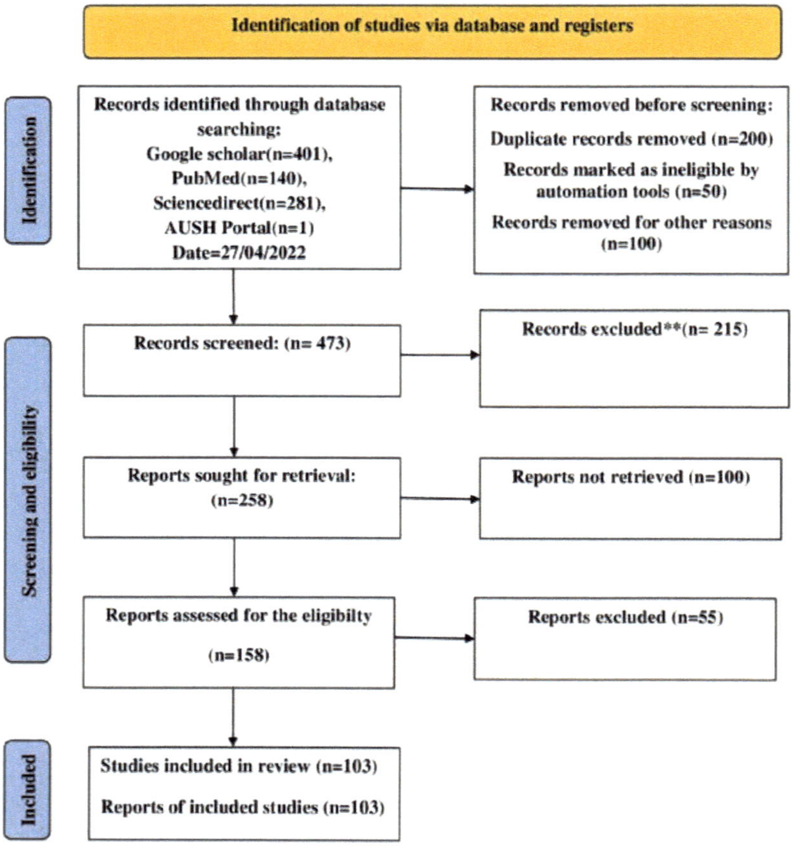


**FIGURE 3.2** Search methodology adopted for the present review.

*Source*: Adapted from Ref. [104].

The entire research methodology is divided into three major categories, which include identification, screening, and a detailed study of the topic. Various databases such as Google Scholar, PubMed, Science Direct, etc., are analyzed to identify a large number of research articles. The use of a reference manager makes it easy to remove duplicate research articles, which are usually eliminated before the screening process. The first step of screening involves the removal of irrelevant articles, which facilitates the sorting of articles, followed by the removal of non-English literature. Articles with insufficient data for this study are eliminated. Finally, this review includes several investigations. The total number of articles is denoted by 'n.'

## 3.4 RESULT AND DISCUSSION

In this literature survey, we have compiled only 38 plants that are well-explored and scientifically validated. Apart from these antidepressant plants, other plants are also reported to have such activity, such as *Boophone distica, Cimicifuga racemosa, Gastrodia elata, Lepidium meyenii,* and *Gentian kochiana,* etc. Attempts to highlight these plants are discussed in Table 3.1 along with their active components and other associated bioactivity.

### *3.4.1 MIMOSA PUDICA*

- **Family:**Fabaceae.
- **Common Name:** Touch-me-not, laajvanti, and Chui-Mui.
- **Description:***Touch-me-not* is a crawling weed. It was mainly recognized as "lajjalu" in Rasayana and has been discovered to have pain-relieving, aphrodisiac, antiasthmatic, and antidepressant properties [3]. *M. pudica* is a traditionally important herb with extensive pharmacological properties, which mainly include antidepressant and anxiety-relieving activity [85]. To treat depression, an aqueous solution of extract from desiccated leaves of Touch-me-not is utilized. The behavioral effects of the extract of Touch-me-not at different doses are investigated in several studies. The findings of one such study suggested that Desipramine (2.14 mg/kg), Clomipramine (1.25 mg/kg), and Touch-me-not (6.0 mg/kg and 8.0 mg/kg) decreased stillness in the forced swimming test and expanded the rate at which reinforcers were received in the DRL-72s test [60].

### *3.4.2 RHODIOLA ROSEA*

- **Family:**Crassulaceae.
- **Common Name:** Rose root, golden root.
- **Description:** *Rhodiola rosea* (*R. rosea*) is a plant adaptogen having antidepressant and anti-stress qualities [7]. It contains a diversity of biological complexes, mainly flavonoids, glycosides, and phenolic

acids, among others [52]. Due to its adaptogenic qualities, the catecholaminergic system enhancer impacts biogenic monoamine levels and inhibits enzymes that enable monoamine breakdown, facilitating neurotransmitter transit in the brain [97]. It also affects the actions of the opioid system and stimulates the generation of opioid peptides, triggering central and peripheral receptors, resulting in an antidepressant impact on the body [82].

### 3.4.3 *CURCUMA LONGA*

- **Family:**Zingiberaceae.
- **Common Name:** Turmeric.
- **Description:***Curcuma longa* has several medical uses, including a protective impact on the brain. Curcumin has the potential to be a long-acting and powerful natural antidepressant [40]. It has been demonstrated in several depressive animal models that curcumin possesses antidepressant qualities. In Forced swim experiments, a curcumin dosage of 10 to 80 mg/kg decreased immobility. In albino mice, curcumin effectively reduced reserpine-induced behavioral sadness at a concentration of 40 to 80 mg/kg [46]. Curcumin has low oral bioavailability. However, when combined with piperine (a bioavailability enhancer), it displays antidepressant-like properties [16].

### 3.4.4 *WITHANIA SOMNIFERA*

- **Family:**Solanaceae.
- **Common Name:** Ashwagandha, poison gooseberry, Indianginseng, and winter cherry.
- **Description:** The medication is made from the plant's roots and is widely used in India. It efficiently manages a wide range of ailments, including illnesses caused by the environment, as well as physical and emotional health. In a six-minute forced swimming exercise, mice given 100 mg/kg of the plant root concentrate displayed decreased mobility [8]. During a similar forced swimming test, 20 and 50 mg/kg p.o. imipramine had antidepressant effects compared to 10 mg/kg imipramine [15].

### 3.4.5 *RHAZYA STRICTA*

- **Family:**Apocynaceae.
- **Common Name:** Senhwar, sahaer, dogbane.
- **Description:** It is ubiquitous throughout Afghanistan and the Arabian Peninsula region and is especially prevalent near oases. The antidepressant properties of Dogbane are attributed to numerous alkaloids and flavonoids found in the aqueous leaf extract, which exhibited antidepressant properties in a forced swimming test [68]. Rats were given different amounts of the freeze-dried extract of the leaves of the plant (0.025–6.4 g/kg) or "desipramine" (10, 20, and 40 mg/kg) before being put through a forced swimming test [6].

### 3.4.6 *GINKGO BILOBA*

- **Family:** Gingkgoaceae.
- **Common Name:** Gingko, maidenhair tree.
- **Description:***Gingko biloba*, a plant native to China can be utilized to heal mental depression lacking the negative side effects associated with manufactured antidepressants [16]. Its interaction with venlafaxine is effective in protecting neurons and minimizing brain damage. After 14 days of oral treatment therapy, the Concentrate of *Gingko biloba* (EGB) significantly reduced the duration of immobility in the FST at doses of 10 and 50 mg/kg body weight, despite no change in motor dysfunction in the open field test (OFT) with similar dosage. These findings suggest that EGB may have antidepressant properties [78].

### 3.4.7 *HYPERICUM PERFORATUM*

- **Family:** Hyperiaceae.
- **Common Name:** St. John's-wort.
- **Description:** The antidepressant properties of the plant are attributed to different flavonoids and alkaloids found in the aqueous leaf extract, which showed antidepressant effects in the forced

swimming test [68]. The primary source of pharmaceutical properties is mainly due to the phloroglucinol derivative "Hyperforin" produced when alcoholic extracts of SJW are used to treat depression. Several studies, however, suggest that flavone derivatives such as "rutin," as well as the "napthodianthrones hypericin" and "Pseudohypericin," play a role in antidepressant effectiveness [99].

### 3.4.8 MORINDA OFFICINALIS

- **Family:**Rubiaceae.
- **Common Name:** Indian mulberry.
- **Description:** The mulberry extract (25 to 50 mg/kg) shows a comparable effect to the antidepressant medicine desipramine (5–10 mg/kg), which significantly reduced the response rate and efficiency ratio while increasing the reinforcement rate. These results provide additional evidence that *Morinda officinalis* extract has antidepressant properties [101].

### 3.4.9 ALLIUM SATIVUM

- **Family:**Liliaceae.
- **Common Name:** Garlic.
- **Description:** When young albino mice of either sex were orally administered an extract of garlic containing alcohol for 14 days at concentrations of 25, 50, and 100 mg/kg, an antidepressant-like effect was observed using the Tail Suspension and Forced Swimming tests [22].

### 3.4.10 ARECA CATECHU

- **Family:** Arecaceae.
- **Common Name:** Betel-nut palm, areca-nut.
- **Description:** The ethanol extract (4–80 mg/kg) of *Areca catechu* significantly reduced immobility time while not affecting spontaneous motor activity [19].

### 3.4.11 *ALOYSIA POLYSTACHYA*

- **Family:** Verbenaceae.
- **Common Name:** Beebrushes, burrito, or burro.
- **Description:** *Aloysia polystachya* is a deciduous shrub, representing a member of the verbena family. When examined in the forced swim test (FST), the hydro-alcoholic leaf extract of *Aloysia polystachya* at concentrations of 12.5, 25, and 50 mg/kg exhibited antidepressant-like effects in Female Sprague-Dawley rats. The primary phytoconstituents responsible for the antidepressant-like activity were thujone and carvone. The extract showed similar effectiveness to imipramine and fluoxetine (10 mg/kg) [23].

### 3.4.12 *CECROPIA GLAZIOUI*

- **Family: Urticaceae.**
- **Common Name:** Yarumo or yagrumo, or more specially Yagrumo hembro ("Female hembra").
- **Description:** *Cecropia glazioui* is a dioecious tree containing various antidepressant properties. The components of butanol, catechin and epicatechin, increased monoamine levels in the hippocampus while inhibiting serotonin, dopamine, and noradrenaline uptake by synaptosomes in various brain regions [72].

### 3.4.13 *CANAVALIA BRASILIENSIS*

- **Family:** Fabaceae.
- **Common Name:** Brazilian Jack-bean.
- **Description:** In the FST, the lectins extracted from plants considerably alleviated the immobility period of "Male Swiss albino mice" [13].

### 3.4.14 *SIPHOCAMPYLUS VERTICILLATUS*

- **Family:** Campanulaceae.
- **Common Name:** Mufumbo.

- **Description:***Siphocampylus verticillatus* is native to the southern part of Brazil. The plant exhibits the presence of different anti-depressant agents, including flavonoids (3-methoxy luteolin), triterpenes (alpha and beta-amirin), and steroids (campesterol and beta-sitosterol). The aqueous alcoholic concentrate has demonstrated antidepressant efficacy in TST and FST tests [73].

### *3.4.15 ANIBA RIPARIA*

- **Family:**Lauraceae.
- **Common Name:** Louro, St. John's wort.
- **Description:** Riparin III is an alkamide-type alkaloid (O-methyl-N-2,6-dihydroxy-benzoyl tyramine) obtained from the fruit of the plant. It exhibited anxiolytic efficacy in mice when evaluated in TST and FST at a dosage of 25 and 50 mg/kg [92].

### *3.4.16 OCIMUM SANCTUM*

- **Family:**Lamiaceae.
- **Common Name:** Tulsi, holy basil.
- **Description:** The concentrate derived from the leaves of *Ocimum sanctum,* commonly referred to as Tulsi or Holy basil plant, exhibits calming and anti-aggressive effects on patients dealing with depression [93]. The effectiveness of the root extract is greater in comparison to the extract obtained from the leaves. The test conducted using the forced swimming and tail suspension methods in mice revealed that the methanolic extract from plant roots showed an antidepressant effect [14]. The ethanolic concentrate of holy basil leaves and the butanol fraction help olfactory bulbectomized rats with depression-like behavior [64].

### *3.4.17 CHRYSTACTINIA MEXICANA*

- **Family:**Asteraceae.
- **Common Name:** Damianita daisy.

- **Description:** Antidepressant effects have been discovered in ferulic acid and caffeic acid. In addition to their anti-inflammatory and antioxidant qualities, ferulic acid is used in the treatment of neuro-degenerative diseases. Apigenin, a flavonoid component found in almost all plants, has antidepressant and anti-anxiety properties. Apigenin demonstrates pharmacological properties by binding to a benzodiazepine GABA receptor [18].

### *3.4.18 MOMORDICA CHARANTIA*

- **Family:** Cucurbitaceae.
- **Common Name:** Karela, bitter gourd, bitter melon, balsam pear, and pare.
- **Description:** It primarily develops in the Amazonian tropical region. Flavonoids, glycosides, steroids, triterpenoids, phenols, alkaloids, and tannins are some of the phytoconstituent compounds found in plants [67]. Stress-induced depression models such as FST and TST are used to assess anxiolytic properties in immature fruit and leaves of Bitter gourd [28, 79].

### *3.4.19 PASSIFLORA FOETIDA*

- **Family:**Passifloraceae.
- **Common Name:** Stinking passionflower.
- **Description:** A herbaceous climber, *Passiflora foetida,* has a long history of usage in Mexican traditional medicine for treating various CNS disorders. It contains chemical ingredients such as hydrocyanic acid, flavonoids, and Harman alkaloids [69]. The bioactive constituent, Harman alkaloids, have been found to exhibit antidepressant activity [21, 91].

### *3.4.20 URTICA DIOCIA*

- **Family:**Urticaceae.
- **Common Name:** Common nettle, stinging nettle, burn nettle, and nettle leaf.

- **Description:** According to the current study, *Urtica dioica* exhibits a distinct antidepressant-like effect in depression model tests. Moreover, the repeated administration of this extract produced a similar impact to that of conventional antidepressants such as fluoxetine and haloperidol [17].

### *3.4.21 ALAFIA MULTIFLORA*

- **Family:**Apocynaceae.
- **Description:** The stem bark contained various phytochemicals such as phenols, tannins, alkaloids, flavonoids, and anthraquinones [55]. Flavonoids and terpenes, derived from plants, may penetrate the blood-brain barrier (BBB) and affect brain function [94]. The flavonoids present in the stem concentrate have high antioxidant properties, suggesting that the plant stem may have pharmacological and antidepressant effects on the central nervous system (CNS) [27].

### *3.4.22 CLITORIA TERNATEA*

- **Family:**Fabaceae.
- **Common Name:** Asian pigeonwing, blue pea, Bluebelline, butterfly pea, Darwin pea, and Cordofan pea.
- **Description:***Clitoria ternatea* is an important ayurvedic herb with numerous mental health benefits [57]. *Clitoria ternatea* is a potential medicinal herb that is used to treat neurological disorders as an essential component of medhyarasayan, a brain tonic [56]. The goal of the research was to investigate the effects of the methanolic extract of Bluebelline on the central nervous system (CNS). It was discovered that the extract possesses anticonvulsant, anxiolytic, antistress, nootropic, and antidepressant properties [30]. In both the TST and FST, the methanolic flower extract of *Clitoria ternatea* significantly reduced immobility time [59]. Phytochemical studies suggest that the roots contain ternatins, alkaloids, flavonoids, saponins, and tannins [34].

### 3.4.23 *BASELLA ALBA*

- **Family:**Basellaceae.
- **Common Name:** Malabar spinach, Indian spinach, vine spinach, Ceylon spinach, climbing spinach, cyclone spinach, Chinese spinach, and east Indian spinach [12, 65, 71, 81, 83].
- **Description:** The primary components found in the plant, including alkaloids, tannins, and flavonoids, were discovered to have similar activity as in FST and TST, but no mechanism was proposed [2, 62].

### 3.4.24 *CAYRATIA JAPONICA*

- **Family:**Vitaceae.
- **Common Name:** Bush killer, Yabugarashi, and Japanese cayratia herb.
- **Description:** Seven flavonoids, such as Apigenin-7-O-beta-D-glucuronopyranoside, luteolin, apigenin, luteolin-7-O-beta-D-glucuronopyranoside, quercetin, and aromadendrine, were derived from the methanolic plant concentrate. Among these, only Quercetin showed a strong repressing response on MAO-A, while luteolin and apigenin showed a weak response [36].

### 3.4.25 *COCHLOSPERMUM RELIGIOSUS*

- **Family:** Bixaceae.
- **Common Name:** Silk-cotton tree, buttercup tree.
- **Description:** In animals, an ethanolic extract of Buttercup tree leaf has anxiolytic and antidepressant properties. The extract had a significant stimulating effect on fluoxetine and imipramine for antidepressant and anxiolytic activities [29].

### 3.4.26 *MELISSA OFFICINALIS*

- **Family:** Lamiaceae.
- **Common Name:** Balm, sweet balm, sweet Mary, bee balm, dropsy plant, honey plant, cure-all.

- **Description:** The Honey plant has been used as a pharmaceutical herb since ancient times due to its numerous biological actions. One of the most notable actions is its effect on the neurological spectrum, including neuroprotective properties, anxiety reduction, and antidepressant effects. The fresh or dried leaves, as well as the uppermost aerial parts of the Honey plant, are consumed for therapeutic purposes [1]. The major active compounds found in *Melissa officinalis* L. are terpenes and phenolics [58]. Rosmarinic acid, Caffeic acid, Luteolin, Naringin, Naringenin, quercetin, and hesperidin are the most significant phenolic compounds present in the leaves and flowers of the Honey plant [86].

### *3.4.27 ROSA DAMASCENE*

- **Family:**Rosaceae.
- **Common Name:** Arab rose, Bulgarian rose, Taif rose, Turkish rose, Ispahan, castile rose, and rose.
- **Description:** It is a decorative plant with medicinal benefits. It also helps alleviate depression, grief, and nervous stress [24]. Arab rose has a wide range of biopharmacological properties, including antidepressant, hypoglycemic, anti-inflammatory, analgesic, antimicrobial, and antioxidant effects. Geraniol and citronellol are the main components of rose essential oil that have pharmaceutical actions [5].

### *3.4.28 AEGLE MARMELOS*

- **Family:**Rutaceae.
- **Common Name:** Indian bael, golden apple, stone apple, Bengal quince, Japanese bitter orange.
- **Description:***Aegle marmelos* have the potential to be used as a natural psychopharmacological agent for anxiety-related disorders like stress and depression. The methanolic leaf concentrate of the stone apple has substantial antidepressant and anxiolytic properties, possibly due to increased monoamine levels at postsynaptic

sites [45]. It is traditionally used as a brain tonic [33]. Additionally, it is a popular medicinal plant with a wide range of pharmacological effects. In behavioral examinations of pain and depression, Stone apple fruit concentrate and aegeline significantly reduced the reserpine-induced decrease in pain tolerance and increased stillness [89].

### 3.4.29 *CUCURBITA PEPO*

- **Family:**Cucurbitaceae.
- **Common Name:** Pumpkin, table queen squash, vegetable marrow, zucchini, and summer squash.
- **Description:** Depression is primarily a health-related issue associated with psychological stress. Pumpkins possess antioxidant, anti-inflammatory, and antidepressant properties [11]. Pumpkin seeds are used to aid in the treatment of depressive disorders due to their L-tryptophan content. Both raw and cooked pumpkin (*Cucurbita maxima*) seeds may help alleviate depression [31].

### 3.4.30 *EPIMEDIUM BREVICORNUM*

- **Family:**Berberidaceae.
- **Common Name:** Fairy wings, barrenwort, bishop's hat, horny goat weed, or Yin Yang Huo.
- **Description:** Several studies have demonstrated the antidepressant properties of water hyacinth extract, which appears to work by silencing Monoamine oxidase A and Monoamine oxidase B while also lowering serum MDA [49, 70, 87, 95, 102]. Flavonoids and alkaloids also exhibit antidepressant properties [70]. The flavonoids derived from *E. brevicornum* increase the levels of neurotransmitters like 5-HT and 5-HIAA, as well as neurotransmitter receptor gene expression in the pituitary [88]. Icarine, a flavonoid compound that can freely pass through the BBB, reduces brain inflammation and oxidative damage, and is the main component of this plant [51].

### 3.4.31 *LAVANDULA OFFICINALIS*

- **Family:**Lamiaceae.
- **Common Name:** Lavender.
- **Description:** A majority of *Lavandula* species have been examined for their antidepressant properties. In mice, aqueous and hydro-alcoholic lavender extracts (100–400 mg/kg) had antidepressant effects and significantly reduced stillness time in the FST and TST, similar to fluoxetine [37]. The antidepressant effects of linalool were inhibited by pre-treatment with WAY100635 and yohimbine, but other compounds did not affect linalool's antidepressant activity [35].

### 3.4.32 *SALVIA SCLAREA*

- **Family:**Lamiaceae.
- **Common Name:** Clary or clary sage.
- **Description:** The antidepressant effect was inhibited by various antagonist receptors, such as dopamine receptors. Therefore, the antidepressant property is shown by the dopaminergic system [84]. An additional study showed that injecting a hydroalcoholic concentrate of clary sage into rats alleviated misery and depression issues [38]. Linalool and apigenin are the antidepressant agents obtained from the plant [9].

### 3.4.33 *CHAMAEMELUM NOBILE*

- **Family:**Asteraceae.
- **Common Name:** Chamomile.
- **Description:** Chamomile has been shown to have antibacterial, antispasmodic, soothing, and anti-inflammatory properties [42]. Compounds isolated from the plants have an antidepressant effect [37]. The main active chemical compounds found in chamomile are flavonoids like quercetin, apigenin, luteolin, and a terpenoid called alpha-bisabolol. Quercetin, the primary ingredient of the plant, has been shown to block monoamine oxidase, resulting in sedative effects [75].

### 3.4.34 *SCHINUS MOLE*

- **Family:**Anacardiaceae.
- **Common Name:** Pink peppercorns.
- **Description:** Dopaminergic, serotonergic, and noradrenergic systems are all involved in the antidepressant effects of n-hexane pink peppercorn extract. In mice, rutin, a flavonoid molecule derived from pink peppercorns, had an antidepressant effect at a concentration of 0.3–3 mg/kg. Rutin is thought to have antidepressant properties by boosting noradrenaline and serotonin in the synaptic gap [53].

### 3.4.35 *BERBERIS ARISTATE*

- **Family:**Berberidaceae.
- **Common Name:** Indian barberry, "Chutro" or "Sumbal" or tree turmeric.
- **Description:** The plant contains Berberine, an isoquinoline alkaloid that acts as an antidepressant in mice when evaluated in FST due to increased dopamine, serotonin, and noradrenaline levels in the brain [47]. Motor activity in mice elevates saddened mood by lowering stillness duration in forced swim and TSTs, which indicates that *Berberis aristate* possesses antidepressant properties [76].

### 3.4.36 *MACROTYLOMA UNIFLORUM*

- **Family:**Fabaceae.
- **Common Name:** Horse gram, Kulthi bean, Hurali, Murali gram.
- **Description:** Daidzein and genistein were found in a grain ethanol extract of horse gram. Wistar rats weighing 150–180 grams of either sex were utilized, and two antidepressant activities were discovered using models FST and TST, as well as potentiation of nor-epinephrine toxicity. The ethanol extracts have antidepressant properties due to their antioxidant and tyrosine kinase inhibitory properties [103].

### 3.4.37 *ZINGIBER OFFICINALE*

- **Family:**Zingiberaceae.
- **Common Name:** Ginger.
- **Description:** Gingerol and shogaol have a strong affinity for LEU 453, PHE 454, and TYR 457. A docking investigation of gingerol and shogaol with commonly used drugs for depression treatment, such as imipramine, revealed that these compounds could be an alternative source for depression treatment. As a result, this strategy is useful for medication discovery and depression therapy [41]. *Zingiber officinale* rhizome phenolic compounds have significant antioxidant, tyrosinase inhibitory, antidepressant, antifungal, and anticancer effects [61].

### 3.4.38 *TINOSPORA CORDIFOLIA*

- **Family:** Menispermaceae.
- **Common Name:** Gurjo, heart-leaved moonseed, Guduchi or Giloy.
- **Description:** Heart-leaved moonseed is widely used in Ayurvedic treatment as a tonic or rejuvenator [63]. Nitric oxide is one of the smallest physiologically active molecules, and it is essential for intracellular signaling. The plant plays a significant role in patients with depression [96].

A list of antidepressant agents derived from various botanical parts of medicinal plants and their reported pharmacological activities are detailed in Table 3.1.

## 3.5 SUMMARY

India has a diverse range of medicinal plants found across the country in various geographical and biodiversity hotspots. Plants have been used to treat a wide range of psychological disorders since ancient times. The use of plant-based antidepressant agents has proven to be an effective alternative to synthetic antidepressants, as they have fewer side effects. This review compiles various plants containing antidepressant agents from various research papers. It can be concluded that there are still many plants that need to be studied for their medicinal benefits, as they are completely risk-free with very few side effects, unlike synthetic medications. Therefore, herbal products should be preferred over synthetic medicines to treat disorders such as stress, depression, and various mental health-related problems.

## ACKNOWLEDGMENT

The authors express their gratitude to the Central University of Himachal Pradesh for providing the necessary infrastructure to conduct these studies. Kirti Raina acknowledges the financial assistance provided by the Central University of Himachal Pradesh for her Doctoral program.

## KEYWORDS

- **antidepressant agents**
- **antidepressants**
- **dopamine**
- **flowering plants**
- **Himalayan flowering plants**
- **neurotransmitters**
- **norepinephrine**
- **serotonin**
- **stress**

## REFERENCES

1. Abdelhalim, A., Karim, N., Chebib, M., Aburjai, T., Khan, I., Johnston, G. A., & Hanrahan, J., (2015). Antidepressant, anxiolytic and antinociceptive activities of constituents from *Rosmarinus officinalis*. *Journal of Pharmacy & Pharmaceutical Sciences*, *18*(4), 448–459.
2. Abhinayani, G., Goud, G. N., Nagamani, K. C., & Kaur, D., (2016). Antidepressant and skeletal muscle relaxant activity of methanolic extracts of *Basella alba* L. *Asian Journal of Biomedical and Pharmaceutical Sciences*, *6*(55), 07–10.
3. Ahmad, H., Sehgal, S., Mishra, A., & Gupta, R., (2012). *Mimosa pudica* L. (Laajvanti): An overview. *Pharmacognosy Reviews*, *6*(12), 115.
4. Ahmed, H. M., (2018). Ethnomedicinal, phytochemical and pharmacological investigations of *Perilla frutescens* (L.) Britt. *Molecules*, *24*(1), 102.

5. Akram, M., Riaz, M., Munir, N., Akhter, N., Zafar, S., Jabeen, F., & Said, K. F., (2020). Chemical constituents, experimental and clinical pharmacology of *Rosa damascena:* A literature review. *Journal of Pharmacy and Pharmacology*, *72*(2), 161–174.
6. Ali, B. H., Bashir, A. K., & Tanira, M. O. M., (1998). The effect of *Rhazya stricta* Decne, a traditional medicinal plant, on the forced swimming test in rats. *Pharmacology Biochemistry and Behavior*, *59*(2), 547–550.
7. Amsterdam, J. D., & Panossian, A. G., (2016). *Rhodiola rosea L.* as a putative botanical antidepressant. *Phytomedicine*, *23*(7), 772–783.
8. Archana, R., & Namasivayam, A., (1999). Antistressor effect of *Withania somnifera. Journal on Ethnopharmacology*, *64*(1), 91–93.
9. Avallone, R., Zanoli, P., Puia, G., Kleinschnitz, M., Schreier, P., & Baraldi, M., (2000). Pharmacological profile of apigenin, a flavonoid isolated from *Matricaria chamomilla. Biochemical Pharmacology*, *59*(11), 1387–1394.
10. Bada, V. G. S., Bhatt, S., Kutagulla, V. K., Kanala, S. R., Yiragam, R. P. R., & Peraman, R., (2022). Antidepressant and anxiolytic activities of *Cochlospermum religiosum* leaf extract, synergism with antidepressants, and molecular docking studies. *Indian Journal of Natural Products and Resources (IJNPR) (Formerly Natural Product Radiance (NPR), 13*(1), 23–31.
11. Balgoon, M. J., Al-Zahrani, M. H., Jaouni, S. A., & Ayuob, N., (2021). Combined oral and topical application of pumpkin (*Cucurbita pepo* L.) alleviates contact dermatitis associated with depression through downregulation pro-inflammatory cytokines. *Frontiers in Pharmacology*, *12*, 663417.
12. Bamidele, O., Akinnuga, A. M., Olorunfemi, J. O., Odetola, O. A., Oparaji, C. K., & Ezeigbo, N., (2010). Effects of aqueous extract of *Basella alba* leaves on haematological and biochemical parameters in albino rats. *African Journal of Biotechnology*, *9*(41), 6952–6955.
13. Barauna, S. C., Kaster, M. P., Heckert, B. T., Do Nascimento, K. S., Rossi, F. M., Teixeira, E. H., & Leal, R. B., (2006). Antidepressant-like effect of lectin from *Canavalia brasiliensis* (ConBr) administered centrally in mice. *Pharmacology Biochemistry and Behavior*, *85*(1), 160–169.
14. Bathala, L. R., Rao, Ch. V., Manjunath, S. M., Vinuta, S., & Vemulapalli, R., (2012). Efficacy of *Ocimum sanctum* for relieving stress: A preclinical study. *The Journal of Contemporary Dental Practice*, *13*(6), 782–786.
15. Bhattacharya, S. K., Bhattacharya, A., Sairam, K., & Ghosal, S., (2000). Anxiolytic-antidepressant activity of *Withania somnifera* glycowithanolides: An experimental study. *Phytomedicine, 7*(5), 463–469.
16. Bhutani, M. K., Bishnoi, M., & Kulkarni, S. K., (2009). The anti-depressant-like effect of curcumin and its combination with piperine in unpredictable chronic stress-induced behavioral, biochemical, and neurochemical changes. *Pharmacological and Biochemical Behavior*, *92*(1), 39–43.
17. Chahardehi, A. M., Ibrahim, D., Abolhassani, F., & Sulaiman, S. F., (2012). Antidepressant-like effect of extracts from *Urtica dioica* in mice model of depression. In: *Proceedings of The Annual International Conference, Syiah Kuala University-Life Sciences & Engineering Chapter* (Vol. 2, No. 1).
18. Colares, M., Muguerza, A., Rosella, M. A., & Consolini, A. E., (2013). Antispasmodic effects of *Mikania micrantha* Kunth and dual gastrointestinal effect of *Mikania*

*cordifolia* (LF) Willd (*Asteraceae*) on isolated rat thin intestine. *Pharmacology Online*, *2*, 1–11.
19. Dar, A., & Khatoon, S., (1997). Antidepressant effects of ethanol extract of *Areca catechu* in rodents. *Phytotherapy Research, 11*(2), 174–176.
20. Dhamija, H. K., Parashar, B., & Singh, J., (2011). The anti-depression potential of herbal drugs: An overview. *Journal of Chemical Pharmaceutical Research, 3*(5), 725–735.
21. Dhawan, K., Kumar, S., & Sharma, A., (2001). Anti-anxiety studies on extracts of *Passiflora incarnate Linneaus. Journal on Ethnopharmacology, 78*(2), 165–170.
22. Dhingra, D., & Kumar, V., (2008). Evidence for the involvement of monoaminergic and GABAergic systems in the antidepressant-like activity of garlic extract in mice. *Indian Journal of Pharmacology, 40*(4), 175–179.
23. Diaz-Veliz, S. M. G., Millan, R., Lungenstrass, H., Quiros, S., Coto-Morales, T., & Hellion- Ibarrola, M. C., (2005). Anxiolytic and antidepressant-like effects of the hydroalcoholic extract from *Aloysia polystachya* in rats. *Pharmacology Biochemistry and Behavior*, *82*(2), 373–378.
24. Dolati, K., Rakhshandeh, H., & Shafei, M. N., (2011). Antidepressant-like effect of aqueous extract from *Rosa damascena* in mice. *Avicenna Journal of Phytomedicine*, *1*(2), 91–97.
25. Eloziia, N., Kumar, N., Kothiyal, P., Deka, P., & Nayak, B. K., (2017). A review on antidepressant plants. *Journal of Pharmacy Research, 11*(5), 382–396.
26. Ferrari, A. J., Charlson, F. J., Norman, R. E., Patten, S. B., Freedman, G., Murray, C. J., & Whiteford, H. A., (2013). Burden of depressive disorders by country, sex, age, and year: Findings from the global burden of disease study 2010. *PLoS Medicine*, *10*(11), e1001547.
27. Foyet, H. S., Tsala, D. E., Bouba, A. A., & Hritcu, L., (2012). Anxiolytic and antidepressant-like effects of the aqueous extract of *Alafia multiflora* stem barks in rodents. *Advances in Pharmacological Sciences*, 1–8.
28. Ganesan, A., Natesan, S., Perumal, P. G., Vellayutham, R., Manickam, K., & Ramasamy, N., (2008). Anxiolytic, antidepressant, and anti-inflammatory activities of methanol extract of *Momordica charantia Linn.* leaves (*Cucurbitaceae*). *Iranian Journal of Pharmacology and Therapeutics*, *7*(1), 43–47.
29. Gari, S. B. V., Bhatt, S., Kutagulla, V. K., & Reddy, K. S., (2022). Antidepressant and anxiolytic activities of *Cochlospermum religiosum* leaf extract, synergism with antidepressants, and molecular docking studies. *Indian Journal of Natural Products and Resources, 13*(1), 23–31.
30. Gautam, R. K., Dixit, P. K., & Mittal, S., (2013). Herbal sources of antidepressant potential: A review. *International Journal of Pharmaceutical Sciences Review and Research*, *18*(1), 86–91.
31. George, S., & Nazni, P., (2012). Antidepressive activity of processed pumpkin (*Cucurbita maxima*) seeds on rats. *International Journal of Pharma Medicine and Biological Sciences, 1*(2), 225–231.
32. Ghasemi, T., Abnous, K., Vahdati, F., Mehri, S., Razavi, B. M., & Hosseinzadeh, H., (2015). Antidepressant effect of *Crocus sativus* aqueous extract and its effect on CREB, BDNF, and VGF transcript and protein levels in rat hippocampus. *Drug Research*, *65*(7), 337–343.

33. Ghosh, S., Kumar, A., Sachan, N., & Chandra, P., (2022). Evaluation of the antidepressant-like effect of total sterols fraction and stigmasterol isolated from leaves of *Aegle marmelos* and possible mechanism (s) of action involved. *Current Drug Discovery Technologies*, *19*(2), 26–40.
34. Gupta, G. K., Chahal, J., & Bhatia, M., (2010). *Clitoria ternatea* (L.): Old and new aspects. *Journal on Pharmacological Research*, *33*(11), 2610–2614.
35. Guzmán-Gutiérrez, S. L., Bonilla-Jaime, H., Gómez-Cansino, R., & Reyes-Chilpa, R., (2015). Linalool and β-pinene exert their antidepressant-like activity through the monoaminergic pathway. *Life Sciences*, *128*, 24–29.
36. Han, X. H., Hong, S. S., Hwang, J. S., Lee, M. K., Hwang, B. Y., & Ro, J. S., (2007). Monoamine oxidase inhibitory components from *Cayratia japonica*. *Archives of Pharmacal Research*, *30*(1), 13–17.
37. Harati, E., Roodsari, H. R. S., Seifi, B., Kamalinejad, M., & Nikseresht, S., (2014). The effect of oral *Matricaria chamomilla* extract and selenium on postpartum depression and plasma oxidant-anti-oxidant system in mice. *Tehran University Medicine Journal*, *71*, 625–634.
38. Herrera-Ruiz, M., García-Beltrán, Y., Mora, S., Díaz-Véliz, G., Viana, G. S., Tortoriello, J., & Ramírez, G., (2006). Antidepressant and anxiolytic effects of hydroalcoholic extract from *Salvia elegans*. *Journal on Ethnopharmacology*, *107*, 53–58.
39. Herrera-Ruiz, M., García-Beltrán, Y., Sergio, M., Díaz-Véliz, G., Glauce, S. B. V., Jaime, T., & Guillermo, R., (2006). Antidepressant and anxiolytic effects of hydroalcoholic extract from *Salvia elegans*. *Journal of Ethnopharmacology*, *107*(1), 53–58.
40. Hurley, L. L., Akinfiresoye, L., Nwulia, E., Kamiya, A., Kulkarni, A. A., & Tizabi, Y., (2013). Antidepressant-like effects of curcumin in WKY rat model of depression is associated with an increase in hippocampal BDNF. *Behavioural Brain Research*, *239*, 27–30.
41. Ittiyavirah, S. P., & Paul, M., (2013). In silico docking analysis of constituents of *Zingiber officinale* as an antidepressant. *Journal of Pharmacognosy and Phytotherapy*, *5*(6), 101–105.
42. Jäger, A. K., & Saaby, L., (2011). Flavonoids and the CNS. *Molecules*, *16*(2), 1471–1485.
43. Jawaid, T., Gupta, R., & Siddiqui, Z. A., (2011). A review on herbal plants showing antidepressant activity. *International Journal of Pharmaceutical Sciences and Research*, *2*(12), 3051.
44. Ji, W., Li, R., Li, M., Wang, S., Zhang, X., Niu, X., Li, Wet al., (2014). Antidepressant-like effect of essential oil of *Perilla frutescens* in a chronic, unpredictable, mild stress-induced depression model mice. *Chinese Journal on Natural Medicines*, *12*(10), 753–759.
45. Kothari, S., Minda, M., & Tonpay, S. D., (2010). Anxiolytic and antidepressant activities of methanol extract of *Aegle marmelos* leaves in mice. *Indian Journal of Physiology and Pharmacology*, *54*(4), 318–328.
46. Kulkarni, S. K., Bhutani, M. K., & Bishnoi, M., (2008). Antidepressant activity of curcumin: Involvement of serotonin and dopamine system. *Psychopharmacology*, *201*(3), 435–442.

47. Kulkarni, S. K., & Dhir, A., (2007). Possible involvement of L-arginine-nitric oxide (NO)-cyclic guanosine monophosphate (cGMP) signaling pathway in the antidepressant activity of berberine chloride. *European Journal on Pharmacology, 569*, 77–83.
48. Kumar, V., Singh, R. K., Jaiswal, A. K., Bhattacharya, S. K., & Acharya, S. B., (2000). Anxiolytic activity of Indian *Abies pindrow* Royle leaves in rodents: An experimental study. *Indian Journal of Experimental Biology, 38*(4), 343–346.
49. Lai, Z., (2001). Clinical application *of Epimedium brevicornum. Journal of Liaoning College of Traditional Chinese Medicine, 3*, 266–273.
50. Lee, S. A., Hong, S. S., Han, X. H., Hwang, J. S., Oh, G. J., Lee, K. S., & Ro, J. S., (2005). Piperine from the fruits of *Piper longum* with inhibitory effect on monoamine oxidase and antidepressant-like activity. *Chemical and Pharmaceutical Bulletin, 53*(7), 832–835.
51. Liu, B., Xu, C., Wu, X., Liu, F., Du, Y., Sun, J., Tao, J., & Dong, J., (2015). Icariin exerts an antidepressant effect in an unpredictable chronic mild stress model of depression in rats and is associated with the regulation of hippocampal neuroinflammation. *Neuroscience, 294*, 193–205.
52. Ma, G. P., Zheng, Q., Xu, M. B., Zhou, X. L., Lu, L., Li, Z. X., & Zheng, G. Q., (2018). *Rhodiola rosea L.* improves learning and memory function: Preclinical evidence and possible mechanisms. *Front Pharmacology, 4*(9), 1415.
53. Machado, D. G., Bettio, L. E. B., Cunha, M. P., Santos, A. R. S., Pizzolatti, M. G., & Brighente, I. M. C., (2008). Antidepressant-like effect of rutin isolated from the ethanolic extract from *Schinus mole L.* in mice: Evidence for the involvement of the serotonergic and noradrenergic systems. *European Journal on Pharmacology, 587*, 163–168.
54. Majeed, H., Bokhari, T. Z., Sherwani, S. K., Younis, U., Shah, M. H., & Khaliq, B., (2013). An overview of biological, phytochemical, and pharmacological values of *Abies pindrow. Journal of Pharmacological and Phytochemicals, 2*(3).
55. Manalachioaie, R., Sevastre, B., Iulia, P., et al., (2010). Comparative evaluation of antidepressant effects of two *Hypericum* Species (*H. perforatum L.* and *H. maculatum C*) in Swiss mice. *Veterinary Medicine, 67*(1), 115–119.
56. Margret, A. A., Begum, T. N., Parthasarathy, S., & Suvaithenamudhan, S., (2015). A strategy to employ *Clitoria ternatea* as a prospective brain drug confronting monoamine oxidase (MAO) against neurodegenerative diseases and depression. *Natural Products and Bioprospecting, 5*(6), 293–306.
57. Margret, A. A., Dhayabaran, V., Suvaithenamudhan, S., & Parthasarathy, S., (2019). Analyzing the antidepressant and drug efflux competence of *Clitoria ternatea L.* as a P-glycoprotein inhibitor to facilitate the blood-brain barrier. *Acta Scientiarum. Biological Sciences, 41*, 46629.
58. Mencherini, T., Picerno, P., Scesa, C., & Aquino, R., (2007). Triterpene, antioxidant, and antimicrobial compounds from *Melissa officinalis. Journal on Natural Products, 70*, 1889–1894.
59. Mittal, S., Gupta, P., & Nigam, V., (2021). Evaluation of antidepressant-like effect of *Clitoria ternatea Linn. Research Journal of Pharmacy and Technology, 14*(12), 6437–6441.

60. Molina, M., Contreras, C. M., & Tellez-Alcantara, P., (1999). *Mimosa pudica* may possess antidepressant actions in the rat. *Phytomedicine, 6*(5), 319–323.
61. Moorkoth, S., Prathyusha, N. S., Manandhar, S., Xue, Y., Sankhe, R., Pai, K. S. R., & Kumar, N., (2021). Antidepressant-like effect of dehydrozingerone from *Zingiber officinale* by elevating monoamines in the brain: In silico and *in vivo* studies. *Pharmacological Reports, 73*(5), 1273–1286.
62. Moragrega, I., & Ríos, J. L., (2021). Medicinal plants in the treatment of depression: Evidence from preclinical studies. *Planta Medica, 87*(9), 656–685.
63. Mutalik, M., & Mutalik, M., (2011). *Tinospora cordifolia:* Role in depression, cognition, and memory. *Australian Journal of Medical Herbalism, 23*(4), 168–173.
64. Nguyen, T. H., Le, T. X., Pham, T. N. H., Nguyen, M. K., & Kinzo, M., (2021). The antidepressant-like effects of an n-butanol fraction of *Ocimum sanctum Linn.* extract in unpredictable chronic mild stress-induced depression in mice. *Vietnam Journal of Science, Technology, and Engineering, 63*(4).
65. Nirmala, A., Saroja, S., & Devi, G. G., (2011). Antidiabetic activity of *Basella rubra* and its relationship: With the antioxidant property. *British Biotechnology Journal, 1*(1), 1–9.
66. Oshima, Y., Matsuoka, S., & Ohizumi, Y., (1995). Antidepressant principles of *Valeriana fauriei* roots. *Chemical and Pharmaceutical Bulletin, 43*(1), 169–170.
67. Patil, S. A., & Patil, S. B., (2011). Toxicological studies of *Momordica charantia Linn.* Seed extracts in male mice. *International Journal on Morphology, 29*(4), 1212–1218.
68. Perviz, S., Khan, H., & Pervaiz, A., (2016). Plant alkaloids as an emerging therapeutic alternative for the treatment of depression. *Frontiers in Pharmacology, 7*, 28.
69. Pongpan, N., Luanratana, O., & Suntornsuk, L., (2007). Rapid reversed-phase high-performance liquid chromatography for vitexin analysis and fingerprint of *Passiflora foetida. Current Science, 93*(3), 378–382.
70. Rabiei, Z., & Rabiei, S., (2017). A review on antidepressant effect of medicinal plants. *Bangladesh Journal of Pharmacology, 1*(1), 1–11.
71. Rahman, M. R., Ali, M., Sharif, M., & Tajmim, A., (2017). A review study on the traditional plants has potential antidepressant properties. *MedCrave Online Journal of Cell Science and Report, 4*(5), 138–145.
72. Rocha, F. F., Lima-Landman, M. T. R., Souccar, C., Tanae, M. M., De Lima, T. C. M., & Lapa, A. J., (2007). Antidepressant-like effect of *Cecropia glazioui* Sneth and its constituents – *In vivo* and *in vitro* characterization of the underlying mechanism. *Phytomedicine, 14*, 396–402.
73. Rodrigues, A. L., Da Silva, G. I., Masteussi, A. S., Fernandes, E. S., Miguel, O. G., Yunes, R. A., Calixto, J. B., & Santos, A. R. S., (2002). Involvement of monoaminergic system in the antidepressant-like effect of the hydroalcoholic extract of *Siphocampylus verticillatus. Life Sciences, 70*, 1347–1358.
74. Rosenzweig-Lipson, S., Beyer, C. E., Hughes, Z. A., Khawaja, X., Rajarao, S. J., Malberg, J. E., & Schechter, L. E., (2007). Differentiating antidepressants of the future: Efficacy and safety. *Pharmacology & Therapeutics, 113*(1), 134–153.
75. Saaby, L., Rasmussen, H. B., & Jager, A. K., (2009). MAO-A inhibitory activity of quercetin from *Calluna vulgaris* (L.) *Hull Journal on Ethnopharacology, 121*, 178–181.

76. Sachan, N., Shrivastav, A., & Pal, D., (2019). Evaluation of the antidepressant activity of ethanolic extract of *Abies webbiana* and *Berberis aristata* in laboratory animals. *Journal of Drug Delivery and Therapeutics*, *9*(1), 244–247.
77. Sairam, K., Dorababu, M., Goel, R. K., & Bhattacharya, S. K., (2002). Antidepressant activity of standardized extract of *Bacopa monniera* in experimental models of depression in rats. *Phytomedicine*, *9*(3), 207–211.
78. Sakakibara, H., Ishida, K., Grundmann, O., Nakajima, J. I., Seo, S., Butterweck, V., & Terao, J., (2006). The antidepressant effect of extracts from *Ginkgo biloba* leaves in behavioral models. *Biological and Pharmaceutical Bulletin*, *29*(8), 1767–1770.
79. Saldanha, T., Kaspate, D., Karmarkar, B., et al., (2015). Evaluation of antidepressant activity of ethanolic extract *Momordica charantia* unripen fruit. *International Journal of Pharmaceutical Sciences and Drug Research*, *7*(1), 68–71.
80. Santosh, P., Venugopal, R., Nilakash, A. S., Kunjbihari, S., & Mangala, L., (2013). Antidepressant activity of methanolic extract of *Passiflora foetida* leaves in mice. *International Journal of Pharmacy and Pharmaceutical Sciences*, 112–115.
81. Saroj, V., Rao, P. S., Rao, S. K., et al., (2012). Pharmacognostical study of *Basella alba* stem. *International Journal of Pharma and Bio Sciences*, *3*(3), 1093–1094.
82. Sarris, J., (2007). Herbal medicines in the treatment of psychiatric disorders: A systematic review. *Phytotherapy Research*, *21*(8), 703–716.
83. Sen, K., Goel, A., Rawal, S., et al., *(*2010*)*. Anti-microbial activity of *Basella rubra* leaves. *International Journal on Pharmacological Science and Research*, *1*(2), 88–91.
84. Seol, G. H., Shim, H. S., Kim, P. J., Moon, H. K., Lee, K. H., Shim, I., Suh, S. H., & Min, S. S., (2010). Antidepressant-like effect of *Salvia sclarea* is explained by modulation of dopamine activities in rats. *Journal on Ethnopharmcology*, *130*(1), 187–190.
85. Shaikh, Z., Roy, S. P., Patel, P., & Gohil, K., (2016). Medicinal value of *Mimosa pudica* as an anxiolytic and antidepressant: A comprehensive review. *World Journal of Pharmacy and Pharmaceutical Sciences*, *5*(3), 430–432.
86. Shakeri, A., Sahebkar, A., & Javadi, B., (2016). *Melissa officinalis L.* – A review of its traditional uses, phytochemistry, and pharmacology, *Journal on Ethnopharmacology*, *188*, 204–228.
87. Shen, Z., Chen, Y., Huang, J., & Hu, Z., (2004). The gene expression profile in hypothalamus-pituitary-adrenal-thymus (HPAT) axis from EF-treated old rats. *Zhongguo Mian yi xue za zhi= Chinese Journal of Immunology*, *20*(1), 59–62.
88. Singh, J. N., Sunil, K., & Rana, A. C., (2013). Antidepressant activity of methanolic extract of *Foeniculum vulgare* (Fennel) fruits in experimental animal models. *Journal of Applied Pharmaceutical Science*, *3*(9), 065–070.
89. Singh, A. P., Singh, L., Singh, P., & Bhatti, R., (2021). Biological evaluation of *Aegle marmelos* fruit extract and isolated aegeline in alleviating pain–depression dyad: In silico analysis of aegeline on MAO-A and iNOS. *ACS Omega*, *6*(3), 2034–2044.
90. Snitz, B. E., Saxton, J., Lopez, O. L., Ives, D. G., Dunn, L. O., Rapp, S. R., Carlson, M. C., et al., (2009). GEM study investigators. identifying mild cognitive impairment at baseline in the *Ginkgo* evaluation of memory (GEM) study. *Ageing Mental Health, 13*(2), 171–182.

91. Soulimani, R., Younos, C., Jarmouni, S., Bousta, D., Misslin, R., & Mortier, F., (1997). Behavioral effects of *Passiflora incarnata L.* and its indole alkaloid and flavonoid derivatives and maltol in the mouse. *Journal on Ethnopharmacology, 57*(1), 11–20.
92. Sousa, F. C. F., Melo, C. T. V., Monteiro, A. P., Lima, V. T. M., Gutierrez, S. J. C., & Pereira, B. A., (2004). Antianxiety and antidepressant effects of riparin III from *Aniba riparia* (Nees) Me (*Lauraceae*) in mice. *Pharmacology Biochemistry and Behavior, 78*, 27–33.
93. Tabassum, I., Siddiqui, Z. N., & Rizvi, S. J., (2010). Effects of *Ocimum sanctum* and *Camellia sinensis* on stress-induced anxiety and depression in male albino *Rattus norvegicus*. *Indian Journal of Pharmacology, 42*(5), 283–288.
94. Tsala, D. E., Theophile, D., Judith, N., et al., (2007). Screening of *Alafia multiflora* for antibacterial, antiradical activity and LD50 investigation. *International Journal on Pharmacology, 3*(4), 327–333.
95. Uppala, P. K., Kumar, K. A., Patro, S. K., & Krishna, M. B., (2015). Experimental evaluation of antidepressant activity of aqueous and chloroform leaf and shoot extracts of *Eichhornia crassipes Linn.* in mice. *Asian Journal of Research in Pharmaceutical Science, 5*(3), 139–144.
96. Walia, V., (2015). *Tinospora cordifolia* in the treatment of depression. *Pharma Tutor, 3*(12), 32–34.
97. Wang, Y. S., Shen, C. Y., & Jiang, J. G., (2019). Antidepressant active ingredients from herbs and nutraceuticals used in TCM: Pharmacological mechanisms and prospects for drug discovery. *Pharmacological Research, 150*, 104520.
98. Wasnik, U., Singh, V., & Ali, M., (2014). Evaluation of the antidepressant effects *of Phyllanthus amarus* in mice. *International Journal of Pharmaceutical Sciences Review and Research, 6*, 26–29.
99. Wurglics, M., & Schubert-Zsilavecz, M., (2006). *Hypericum perforatum*: A 'modern herbal antidepressant. *Clinical Pharmacokinetics, 45*(5), 449–468.
100. Xu, C., Luo, L., & Tan, R. X., (2004). The anti-depression potential of herbal drugs: An overview. *Journal of Ethnopharmacology, 91*(2, 3), 345–349.
101. Zhang, Z. Q., Yuan, L., Yang, M., Luo, Z. P., & Zhao, Y. M., (2002). The effect of *Morinda officinalis.* How, a Chinese traditional medicinal plant, on the DRL 72-s schedule in rats and the forced swimming test in mice. *Pharmacology Biochemistry and Behavior, 72*(1, 2), 39–43.
102. Zhong, H. B., Pan, Y., & Kong, L. D., (2005). The antidepressant effect of *Epimedium brevicornum* extracts. *Chinese Traditional and Herbal Drugs, 10*, 34–40.
103. Zhu, Y., Li, X., Yang, W., Jia, H., Liu, C., Prashar, Y., & Bais, S., (2018). The antidepressant activity of standardized *Macrotyloma uniflorum* extract in experimental models of depression in rats. *International Journal on Pharmacology, 14*(6), 848–855.
104. Haddaway, N. R., Page, M. J., Pritchard, C. C., & McGuinness, L. A. (2022). PRISMA2020: An R package and Shiny app for producing PRISMA 2020-compliant flow diagrams, with interactivity for optimised digital transparency and Open Synthesis. Campbell Systematic Reviews, 18, e1230. https://doi.org/10.1002/cl2.1230

# PART II

# Antidepression Agents from Himalayan Flowers, Mushrooms, Grains, Beans, and Nuts

## CHAPTER 4

# Antidepressant Activity of Himalayan Roots and Tubers

MUHAMMAD NADEEM AKHTAR,[1] ANEES AHMED KHALIL,[1] ABDUR RAUF,[2] SAIMA NAZ,[3] ABDULLAH S. M. ALJOHANI,[4] and WALEED AL ABDULMONEM[5]

*[1]University Institute of Diet and Nutritional Sciences, Faculty of Allied Health Sciences, The University of Lahore, Pakistan*

*[2]Department of Chemistry, University of Swabi, Swabi, Anbar, Khyber Pakhtunkhwa, Pakistan*

*[3]Institute of Biotechnology and Microbiology, Bacha Khan University, Charsadda, Khyber Pakhtunkhwa, Pakistan*

*[4]Department of Veterinary Medicine, College of Agriculture and Veterinary Medicine, Qassim University, Buraydah, Saudi Arabia*

*[5]Department of Pathology, College of Medicine, Qassim University, Buraydah, Saudi Arabia*

### ABSTRACT

Nowadays, the prevalence and co-morbidities associated with various psychiatric ailments, specifically insomnia, anxiety, and depression, are becoming very common. Depression is a psychiatric problem that is causing various behavioral issues in people across the globe. Traditionally, numerous medicinal and herbal prescriptions have been used for the treatment of depression. However, the side effects associated with the use of antidepressant drugs have raised great concern in the community. As a

Himalayan Medicinal Plants for the Treatment of Depression: A Source of Rich Antidepressant Agents.
Abdur Rauf, Prabhakar Semwal, & Hafiz Ansar Rasul Suleria (Eds.)

result, the use of natural products with potent antidepressant activity has increased. Over the last two decades, various investigations on different natural plants and their parts, including roots and tubers, have gained the attention of scientists. The Himalayan region is considered an abundant source of medicinal plants that possess various bioactivities, such as anticancer, antidepressant, antidiabetic, antihypertensive, and cardio-protective properties. Literature has shown that roots and tubers present in the Himalayan region possess antidepressant potential. In this chapter, we discuss the antidepressant potential of roots and tubers present in the Himalayan region.

## 4.1 INTRODUCTION

Depression is characterized as a "state of low mood and disinterest in activities that affects one's behavior, thoughts, feelings, and sense of well-being." It is defined by fluctuations in a person's mood, lack of interest in their surroundings, melancholia, and psychomotor retardation. The World Health Organization (WHO) has stated that nearly 450 million people are suffering from behavioral and mental disorders [61]. Globally, behavioral and mental disorders account for 12.30% of the global burden and are projected to increase to approximately 15% by the year 2020 [55]. These mental and psychiatric ailments are closely linked to suicides, with an estimated 10–20 million suicide attempts occurring each year. Among other psychiatric illnesses, depression is characterized as the most prevalent mental disorder. Depression is a state in which a person experiences an imbalance of energy, slowed thinking and activity, a clouded mental state, feelings of hopelessness, and a temptation towards suicide. Despite the development and availability of antidepressant drugs such as SSRIs (selective serotonin reuptake inhibitors), MAO-A (selective reversible monoamine oxidase-A inhibitors), and SNRIs (selective noradrenaline reuptake inhibitors), depression continues to be one of the main health concerns worldwide [6].

In developing countries, the usage of traditional medicine has played a vital role in fulfilling basic health requirements and is recognized as an important basis for maintaining good health conditions [65]. Developed communities have also increasingly relied on herbal and natural products for the treatment of anxiety and depression due to their safety and effectiveness. It is estimated that nearly three-quarters of the global population relies on traditional medications in their primary healthcare system [18]. Various investigations have shown that different phytochemicals, such as phenolic acids, alkaloids, flavonoids, tannins, saponins, and terpenoids, demonstrate antidepressant and anxiolytic properties. Additionally, several plant parts, including flowers, stems, roots, tubers, fruits, and/or the whole plant, have been investigated for their antidepressant-like activities. Antidepressant models adopted in different studies include monoamine oxidase inhibition assay, open field test (OFT), tail suspension test (TST), hole board test, forced swim test (FST), sucrose preference test (SPT), chronic unpredictable stress, etc. [41]. Neurological studies have helped us better understand the pathophysiology of the disease and the mechanism of action of formulated therapies. The identification of novel natural biomolecules as potent antidepressant agents has opened a new window of opportunity for the development of effective, safe, reliable, and affordable alternatives to synthetic drugs. Scientists are currently investigating numerous plants and their associated metabolites for the effective management of mental disorders [6].

Diverse plants present in the Himalayan region are considered to be a rich source of medicinal potential, such as anticancer, antidepressant, antidiabetic, antihypertensive, and cardio-protective properties. It is estimated that nearly 1957 plant species are present in the Himalayan region. The Indian Himalayan region comprises about 6,000 plant species, out of which 1,748 have been extensively investigated for their medicinal properties [26, 60]. People residing in the Himalayan region have been benefiting from the medicinal properties of the higher plants in this area. Since ancient times, several plants of the Himalayan belt have been used as curative agents against various medical conditions, including depression. This chapter will focus on Himalayan roots and tubers with antidepressant properties.

## 4.2 ROOTS AND TUBERS AS ANTIDEPRESSANTS

### 4.2.1 *ALLIUM SPECIES*

Himalayan onion, botanically known as *Allium wallichii Kunth,* belongs to the family *Amaryllidaceae.* The Himalayan onion is considered a rare and underutilized species with functional and nutraceutical properties. Due to its associated health benefits, it is consumed by people living in the Himalayan region [43, 68]. Briefly, the Himalayan onion is a perennial plant found in high altitude areas (2,700–3,600 m) of Nepal, India, Myanmar, and Bhutan [9, 15, 22, 54]. Various parts of this plant, including the whole plant, roots, bulb, and leaves, have been traditionally used for treating ailments such as flu, cough, cholera, and diarrhea. The bulb of this plant has been used after boiling and frying to lower serum cholesterol levels and treat dysentery [50, 64].

*Allium cepa* L. is another specie of family Amaryllidaceae has shown potent health promoting benefits. *Allium cepa* L. is commonly known as common onion or bulb onion. This plant is widely distributed across different regions of the world, including the Himalayan region [57]. The antidepressant properties associated with *A. cepa* have been evaluated in various studies. In one study, rats were administered onion-quercetin rich bulb in a behavioral model of depression. The forced swimming test (FST) was used to assess the antidepressant-like effect of onion in the experimental rat model. Experimental rats were administered *Allium cepa* (50 mg $kg^{-1}$ B.W.) for 14 days. The results of the forced swimming test showed that the administration of onion powder significantly decreased the immobility time without causing any change in motor dysfunctionality. After the forced swimming test, the content of corticosterone increased, but the pre-treatment with onion powder had no modulatory effect on this increase. In the rat hypothalamus, the forced swimming test resulted in an elevation of dopaminergic activity, while the administration of onion powder downregulated the enhancement of this neurotransmitter. A similar preventive effect was observed with a high dose of onion powder. The outcomes of this study revealed that onion powder exhibited antidepressant-like potential in the experimental behavioral model, which was independent of the hypothalamic-pituitary-adrenal axis [58].

Likewise, the antidepressant and anxiolytic properties of *A. cepa* have also been studied in various investigations. Specifically, in an experimental mice model, onion powder dissolved in water was administered for 14 days at a dose of 200 mg $kg^{-1}$ per day. Both the control and onion bulb treated mice were further divided into stressed and unstressed groups. The mice in the stressed group were exposed to immobilized stress for a duration of 2 hours. An anxiogenic behavior was observed in the mice subjected to immobilized stress. The results of the FST indicated that

2 hours of immobilization stress led to depression in the experimental mice. However, the behavioral deficits caused by immobilization stress were attenuated by the administration of onion bulb. Maximum memory performance was observed in the immobilized-stress induced mice that were pretreated with onion bulb in the MWM (Morris water maze). The results of the aforementioned study revealed that antioxidative enzymes mitigated the depression and anxiety induced by 2 hours of immobilized stress. Furthermore, the cognitive functionality also improved with the administration of onion bulb. In conclusion, the administration of onion bulb may be beneficial in the treatment of depression and anxiety, as well as the improvement of memory functionality [59].

**FIGURE 4.1** *Allium cepa*

### 4.2.2 *VALERIANA JATAMANSI* (SUGANDHAWAL)

Sugandhawal is the native name of *Valeriana jatamansi*, which is a renowned temperate and sub-tropical herbal plant found in Nepal, Meghalaya (Northeast state), and Uttarakhand (Himalayan region) [8, 53]. Traditionally, it has been used as a predominant source of medicine, cosmetics, and in the pharmaceutical industry since ancient times [16, 30]. This plant, particularly its oil, rhizomes, roots (Figure 4.2), and marc, has been well-known for its medicinal properties [52]. The trade and usage of *Valeriana jatamansi* have increased in the last decade due to its nutraceutical potential and the growing demand from the pharmaceutical industry [62]. *Valeriana jatamansi* rhizomes and roots possess medicinal properties such as promoting digestion, reducing diarrhea, relieving pain, and acting as a sedative [17, 25, 51, 56]. Iridoids, essential oil, alkaloids, and flavonoids are some of the bioactive components present in *Valeriana jatamansi* [37]. Iridoids are known to be the main constituents of *Valeriana jatamansi* with the least toxic effects [73].

Various extracts of *Valeriana jatamansi* roots and rhizomes have shown significant antidepressant potential [66]. Recently, the effect of TIV (total iridoids) present in *Valeriana jatamansi* roots and rhizomes has been assessed for its antidepressant activity. TIV augmented the depression-related disorder and visceral hypersensitivity by reducing gastrointestinal motility and increasing 5-HT and NE. Moreover, treatment with TIV was found to be safe at the experimented dosages as no toxicity was observed in experimented rats and mice. The lethal dosage having LD50 (50%-mortality rate) in experimented mice was stated to be above 2,000 mg $kg^{-1}$ B.W. [73]. The antidepressant potential of TIV is considered to be associated with the regulation of different metabolic pathways and gut microflora [70].

The ethanolic (70%) extract of *Valeriana jatamansi* roots and rhizomes contains a rich source of iridoids. A chronic unpredictable mild stress (CUMS) model was established, in which rats were administered with TIV of *Valeriana jatamansi* roots and rhizomes extract (5.7, 11.40, and 22.80 mg $kg^{-1}$ B.W.) or fluoxetine (2.60 mg $kg^{-1}$ B.W.) for 14 days. The results of this study showed that treatment with TIV decreased tail suspension time (TST) and elevated the SPT results. Administration of iridoids modulated the gut flora, exhibited antidepressant activity, and protected the blood-brain barrier (BBB) [87]. Reduced expression of ZO-1 results in increased permeability of the BBB, while BBB dysfunctionality is caused by dephosphorylation of occludin, which affects the structure of tight junctions (TJs) [44]. CUMS has been reported to affect the expression of occludin and ZO-1, thus altering the BBB structure. On the other hand, the iridoids present in *Valeriana jatamansi* roots upregulated the expression of ZO-1 and occludin in order to decrease the permeability of the BBB in the experimented depressed mice. Both low and high doses of iridoids administered to mice showed improved TJ ultrastructure alterations in the BBB [77].

**FIGURE 4.2** *Valeriana jatamansi* roots.

In Chinese folk medicine, *Valeriana jatamansi* roots have been extensively used due to the presence of iridoids, which are the most abundant bioactive components. The antidepressant potential of iridoids present in *Valeriana jatamansi* roots has been examined by Wang et al. [70]. CUMS mice were administered with varied concentrations (5.7–22.9 mg $kg^{-1}$ B.W.) of Valeriana jatamansi roots for a period of 1 week. The results of this study showed that the CUMS mice model demonstrated significant signs of depression, while the group subjected to fluoxetine (standard drug) and experimented with *Valeriana jatamansi* roots showed augmented depression symptoms in the experimented mice. Moreover, iridoids helped in modulating the content of CRF (corticotropin-releasing factor), NE (norepinephrine), 5-HT (5-hydroxytryptamine), and SP (substance P) in both the colon and hippocampus. In conclusion, the antidepressant potential of iridoids may be linked to the expression of CRF, NE, 5-HT, and SP in the intestine and brain [70]. The main bioactive components present in *Valeriana jatamansi* roots are considered to be responsible for its potential to lower depression. The ethanolic extract and valepotriates present in the roots of *Valeriana jatamansi* have been reported to be potent agents to suppress anxiety and induce hypogenesis [82]. Similarly, the ethanolic extract demonstrated a significant decrease in the levels of NE and 5-HT, as experimented in Sprague Dawley (SD) rats. Ethanolic extract was administrated with different concentrations (0.3–0.9 g $kg^{-1}$ B.W.) daily [76].

In a study conducted by Yan et al. [75], dysfunctionality of the hypothalamus-pituitary-adrenal axis was examined in anxiety model rats. The researchers reported that levels of β-EP (blood β-endorphin) and CS were highest in anxiety rats compared to normal rats. Rats administered with an ethanolic extract showed suppressed expression of the genes Orexin and CRH (corticotropin-releasing hormone) [75]. In traditional Chinese medicine, the rhizomes and roots of *Valeriana jatamansi* are used to create a medicine known as Zhi Zhu Xiang (ZZX). ZZX has historically been used to treat disorders related to depression, sleep, and anxiety. In terms of etiology and clinical manifestations, PTSD (post-traumatic stress disorder) shares similarities with anxiety and depression, and ZZX has been reported to be helpful in preventing and treating PTSD. A study was conducted to assess the potential effects of the ethanolic extract of ZZX against PTSD in a mouse model, using various behavioral examinations. The results showed that the ZZX extract effectively managed PTSD symptoms through the eCB (endocannabinoid), HPA axis, and neurotransmitter system. In the amygdala, prefrontal cortex, and hippocampus of PTSD mice, the levels of glutamic acid, 5-HT, and NE were elevated, while γ-aminobutyric acid (GABA) and dopamine were reduced in these brain regions. Furthermore, the application of the extract restored the expression of these regions. Furthermore, the administration of ZZX extract decreased corticosterone, CRH, and adrenocorticotropic hormone in the serum and brain sections of PTSD mice [78]. The predominant constituent present in different parts of *Valeriana jatamansi,* specifically its roots, is Valtrate. This has been used in the treatment of different mood-related disorders. According to a study, isolated valtrate was used to examine its anxiolytic potential in rats using EPM (elevated plus-maze) and OFT. Isolated valtrate was administered to rats at three different concentrations, i.e., 5, 10, and 20 g $kg^{-1}$ B.W. on a daily basis for a period of 10 days. Valtrate administered at a dose of 10 mg $kg^{-1}$ significantly decreased the levels of corticosterone in the serum of the examined rats [63].

### 4.2.3 MORINDA OFFICINALIS

*Morinda officinalis* How (MO) is a medicinal shrub that is cultivated in several tropical and subtropical since last 200 years. It is a well-known herbal plant in different regions of Asia, such as Pakistan, India, South

China, Nepal, and Bhutan [86]. MO has also been reported to be widely cultivated in hilly areas [27]. Its roots are known as Baji tian, which is the traditional name of *Morinda officinalis* How in Chinese medicine and has been used for the preparation of various tonics and nutritional supplements (Figure 4.3). Traditionally, MO has been used for the treatment of various ailments such as impotence, dermatitis, depression, osteoporosis, Alzheimer's, and rheumatoid arthritis [72, 79, 90]. Various bioactivities of *Morinda officinalis* roots have been documented in the literature, including antiradiation, antidepressant, antirheumatoid, anti-Alzheimer's, antiosteoporosis, antioxidant, anti-inflammatory, cardioprotective, immunomodulatory, antifatigue, and anticancer properties [14, 29, 33, 79, 84, 85, 91]. Phytochemical investigations have shown that the root of *Morinda officinalis* possesses a variety of bioactive ingredients such as essential oil, anthraquinones, oligosaccharides, polysaccharides, and iridoids glucosides [13].

As mentioned above, the roots of MO comprise a rich source of different polysaccharides and oligosaccharides. In a study, polysaccharides present in MO root were administered to a depression model of rats. The results of this study showed a reduction in the number of mistakes in the T-type labyrinth test and elevated levels of SOD (superoxide dismutase) in the brain. Furthermore, the administration of MO polysaccharides elevated the number of neurons and mitigated oxidative stress conditions, resulting in an improvement in cognitive behavior in depressed experimental rats [40]. Chronic stress induced hyperplasia in the adrenal cortex, decreased WBC (white blood cell count), reduced testosterone levels, and elevated corticosterone levels in experimental rats and mice. Intraperitoneal administration of MO-oligosaccharide (MW-97) at a dose of 100 mg $kg^{-1}$ B.W. 15 days before the induction of stress helped in the reversal of the above-stated alterations in experimental animal subjects. The results of this study highlighted the importance of MO-oligosaccharide in the reversal of depression-like symptoms induced in experimental rats and mice without any detrimental effects on the CNS (central nervous system) [36].

MO oligosaccharides administered at a dose of 20 to 50 mg $kg^{-1}$ B.W. decreased immobility time as examined through the forced swimming test. Similarly, in a learned helplessness depressive animal model, MO oligosaccharide administered at a dose of 60 mg $kg^{-1}$ B.W. caused a reduction in the number of failing animals [89]. MO-oligosaccharide resulted in an antidepressant effect and helped in promoting pliability against stress as determined via the forced swimming test. MO-oligosaccharides augmented the antidepressant-like effect by increasing the expression of phosphorylated-Ser 9-glycogen synthase kinase-3β, synaptic proteins, and β-catenin [35, 74]. In inescapable foot shock-induced depression in experimental rats, which was reversed by the administration of MO oligosaccharides (100 mg $kg^{-1}$) [88].

The State Food and Drug Administration (SFDA) of China has approved the use of MO-oligosaccharide for the prevention and treatment of mild to moderate depression. Various clinical trials conducted on MO oligosaccharides have shown that its utilization is effective in treating mild to moderate depressive conditions [86]. Similarly, various experiments, including the head twitches test, have been conducted to determine the pharmacokinetic behavior of MO oligosaccharides. In mice, the administration of MO oligosaccharides resulted in an increase in the number of head twitches, 5-HT, and 5-HIAA, highlighting the antidepressant effect of MO oligosaccharides, which may be attributed to elevated activity of 5-HT. Western blotting analysis showed that MO oligosaccharides significantly increased the expression of synapse proteins and GSK-3β in depressed male rats, indicating that the antidepressant properties of MO oligosaccharides may depend on the regulation of vital neurotrophic pathways. Furthermore, real-time PCR analysis revealed that the administration of MO oligosaccharides for 21 days led to an increase in brain-derived neurotrophic factor in rat nerve growth factor (NGF). They reported that the antidepressant effect of MO oligosaccharides may be due to the protection of corticosterone-induced injured neurons [34].

**FIGURE 4.3** *Morinda officinalis*

### 4.2.4 DAUCUS CAROTA

*Daucus carota* is commonly known as wild carrot and belongs to family Apiaceae. It is cultivated as a root vegetable throughout the world. *Daucus carota* is cultivated around the globe and is thought to be native to Asia (Figure 4.4). It has also been reported to be grown in hilly areas, such as some regions of Nepal [47]. They have been well-known in human nutrition since ancient times due to the presence of ample amounts of bioactive compounds and being a rich source of dietary fiber. Phytochemicals present in *Daucus carota* also play an important role in disease prevention and treatment. *Daucus carota* mainly contains carotenoids, phenolic compounds, ascorbic acid, and polyacetylenes [1]. Additionally, it contains vitamins (A, C, and E) and is also considered a rich source of flavonoids and β-carotene [31]. These bioactive compounds present in *Daucus carota* are responsible for various biological activities, including anticancer, antioxidant, anti-inflammatory, antidepressant, and cardioprotective properties [1].

According to a study, *Daucus carota* has been explored for its potent antidepressant-like activity in a depressive mice model. The Tail Suspension Test (TST), reserpine-induced hypothermia (RIH), Forced Swim Test (FST), 5-HTP potentiation of head twitches in mice (HTPPH), and apomorphine-induced hypothermia (AIH) have been employed to assess the antidepressant properties of *Daucus carota* (DC). For this experimentation, Swiss Albino mice weighing between 25 to 30 grams were used. In AIH and RIH, desipramine at a dose of 20 mg/kg B.W. was used as a standard treatment. While for TST, FST, and HTPPH twitches in mice, fluoxetine at a dose of 25 mg/kg was used as a standard drug. In FST and TST, the duration of immobility was measured. Results of this study revealed that both *Daucus carota* and fluoxetine significantly reduced immobility compared to the control group, as determined through TST and FST. *Daucus carota* showed antidepressant potential when administered at a dose of 400 mg/kg, compared to the standard drugs. Evidently, *Daucus carota* may be helpful as an adjuvant for treating clinical symptoms associated with depression. *Daucus carota* extracted using ethanol has also been found to have antidepressant potential through the inhibition of Serotonin uptake, which behaves as G-protein coupled receptors as mood enhancers. The results of this study focus on validating the traditional use of *Daucus carota* against depression. The ethanolic extract of *Daucus carota* in experimented models revealed an inhibitory effect on Serotonin reuptake, thereby exerting its antidepressant potential [6].

Similarly, the aqueous extract of *Daucus carota* has also been examined for its potent antidepressant properties. This study involved Albino mice weighing 25 to 30 grams. Guidelines suggested by the Organization for Economic Co-operation and Development were followed to assess the toxicity investigations. For toxicity studies, doses of 5–2,000 mg $kg^{-1}$ p.o. were administered to mice and were found to be safe for consumption as no detrimental effects were observed. The aqueous extract of *Daucus carota* was administered to depressive mice at two different concentrations, namely 75 and 150 mg $kg^{-1}$ B.W. Both intervention doses demonstrated significant potential in relieving depression-related conditions and improving the mood of the experimented mice, as examined through the FST, elevated plus maze (EPM) test, and hole board test. Compared to the low experimented dose, the highest dose of *Daucus carota* extract revealed better results in terms of reduced immobility duration and an increased number of rotation counts. In the FST with activity wheel, imipramine was used as a standard treatment. In the hole board test, the maximum administered dose of Daucus carota extract showed a similar potential to that of diazepam (standard drug; 4 mg $kg^{-1}$ B.W.). Even a low dose of *Daucus carota* extract resulted in an increase in head dipping. Additionally, in the EPM test, a high concentration of *Daucus carota* extract was significantly more effective, increasing the number of entries in the open arm and the mean time spent in the open arm, while decreasing the time spent in the closed arm. The findings of this study indicate that *Daucus carota* peel extract has anxiolytic and antidepressant potential. The authors suggest further detailed pharmacological investigation to better understand the mechanisms associated with this bioactivity [45].

The ethanolic extract of *Daucus carota* seeds has a potent effect on brain cholinesterase activity and cognitive functionality, as demonstrated in a mice model. In both young and aged mice, the administration of *Daucus carota* extract at a rate of 200 and 400 mg $kg^{-1}$ B.W. resulted in improved memory scores. The passive avoidance apparatus was used to enhance memory scores in aged mice. *Daucus carota* extract effectively reversed the amnesia induced by the administration of diazepam and scopolamine at rates of 1 mg $kg^{-1}$ and 0.4 mg $kg^{-1}$, respectively. In both young and aged mice, both extracts of *Daucus carota* reduced the acetylcholinesterase (AChE) potential in the brain. The seeds contain choline, which inhibits brain cholinesterase activity and increases acetylcholine levels through elevated acetylcholine synthesis. Therefore, they are helpful in cognitive dysfunctionality [7].

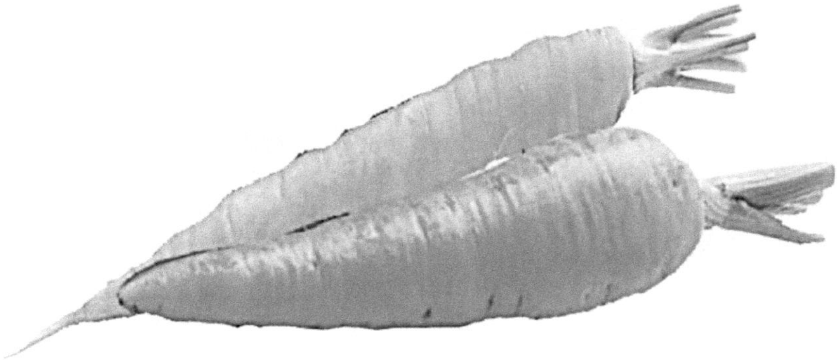

**FIGURE 4.4** *Daucus carota.*

### *4.2.5* DISCOREA ALATA

*Discorea alata* is commonly known as yam and belongs to family *Dioscoreaceae*. Tubers (Figure 4.5) of this plant are single, vary in shape and size, and have underground tuberous roots. *Discorea alata* is grown in Himalayan ranges of Nepal and its tubers are used as an edible source owing to their medicinal properties. The rhizome of this plant has been consumed since ancient times as a remedy for cough and sore throat [19]. Literature has demonstrated that *Discorea alata* has potential as an antidiabetic, antioxidant, antibacterial, and antidiarrheal agent [5]. Furthermore, the neuroprotective potential of *Discorea alata* has been examined in experimental rats under conditions of hyper-homocysteinemia-induced oxidative stress in the brain [69]. Figure 4.5 depicts a visual representation of *Discorea alata* (Yam).

According to a recent study conducted in 2018, the neuropharmacological potential of *Discorea alata* tuber extracted using methanol as an extraction solvent was examined in an experimental mice model. Various standard procedures such as TST, EPM test, force swim test, hole cross test, open-field test, and hole board test were used to assess behavioral changes and activities in the experimental mice. The results of this study revealed that the methanolic extract of *Discorea alata* exhibited significant concentration-dependent potential for neophilia, as assessed through the hole board and EPM tests. The time spent in the open arm was observed to be 28, 82, and 19 seconds for the groups that were fed with 250 mg $kg^{-1}$, 500 mg $kg^{-1}$, and the control group, respectively. Additionally, the crude extract also showed a significant concentration-dependent decrease in exploratory potential due to a reduction in serotonin levels in the experimented mice, as examined using the force swim, tail suspension, hole cross, and OFTs at both administered doses. The proposed inhibitory activity of the extracts on noradrenaline and serotonin had a significant impact on motor activity. In conclusion, the researchers suggested that the methanolic extract of *Discorea alata* may serve as a potent candidate for a natural antidepressant [2].

**FIGURE 4.5** *Discorea alata.*

### 4.2.6 *GASTRODIA ELATA*

*Gastrodia elata* is a well-renowned medicinal herb belonging to family *Orchidaceae*. It has been cultivated in various areas of Bhutan, Nepal, Japan, and China. It is commonly known as Tian-ma in Chinese medicine and has been used traditionally for treatment of different ailments specifically neurological disorders and psychosomatic diseases [11, 12,

83]. Additionally, *Gastrodia elata* is known as a functional ingredient in Taiwan and is utilized to treat issues such as convulsion, dizziness, and headaches. It has been reported to possess anti-inflammatory, antioxidant, anticonvulsant, and antiepileptic properties [20, 21, 23, 39, 46]. Moreover, *Gastrodia elata* has been consumed as a neuroprotective and memory-enhancing agent against ischemic damage, Parkinson's disease, and amyloid β-induced cell death [3, 28, 42]. The results of these studies have highlighted the neurodegenerative potential of *Gastrodia elata*.

According to a study, *Gastrodia elata* extract enhanced symptoms of depression in experimental rats through the reduction of monoamine metabolism and the suppression of the Slit-Robo pathway processes [10, 38]. 4-hydroxybenzyl alcohol (HBA) and gastrodin (GAS) were found to be the bioactive compounds in *Gastrodia elata* extract, and they were thought to have neuroprotective, memory-enhancing, and antiepileptic activities [4, 48, 49, 71, 81]. Previously, *Gastrodia elata* extract was also found to have an ameliorative effect on depression-like symptoms in depressive rats, as determined using FST [10, 38]. *Gastrodia elata* extract, GAS, and HBA showed potent antidepressant potential in experimental rats, and hence it can be concluded that this plant can be used as a potential candidate in the treatment of depression [12]. Orally administered *Gastrodia elata* extract and its components (4-HA and 4-HBA) helped in increasing the time spent percentage and arm entries in the open arms of EPM in experimental mice. The results of this study suggested the anxiolytic-like effects of *Gastrodia elata* extract [24] (Figure 4.6).

## 4.3 SUMMARY

Various individuals in the modern world are facing problems due to different types of psychiatric disorders, particularly depression. The burden and prevalence of depression are increasing day by day, and mental disorders have not only impacted people's daily lives but have also resulted in economic losses for society. In recent decades, scientists have focused on the psychopharmacology of medicinal and herbal plants, driven by the shift in consumer preference towards natural products. Literature suggests that the use of whole plants and/or plant parts (such as roots and tubers) may have complex psychotropic mechanisms associated with the

**FIGURE 4.6** *Gastrodia elata* tuber.

alleviation of depression. However, the pathogenesis and etiology of roots and tubers used for the prevention and treatment of depression are not yet fully understood. Furthermore, clinical trials need to be conducted to gain a better understanding of the mechanisms underlying the antidepressant activities of roots and tubers.

## KEYWORDS


- **allium species**
- **antidepressants**
- **depression**
- **Himalayan roots**
- **psychiatric problems**
- **psychotropic mechanisms**
- **tubers**

## REFERENCES


1. Ahmad, T., Cawood, M., Iqbal, Q., Ariño, A., Batool, A., Tariq, R. M. S., Azam, M., & Akhtar, S., (2019). Phytochemicals in *Daucus carota* and their health benefits. *Foods*, *8*(9), 424.
2. Amin, K. R., Rashid, M. M. O., Uddin, M. G., & Sharmin, T., (2018). New insight in neuropharmacological activities of *Dioscorea alata. Discovery Phytomedicine-Journal of Natural Products Research and Ethnopharmacology*, *5*(1), 1–6.
3. An, H., Kia, I. S., Koppula, S., Kim, B. W., Park, P. J., Lim, B. O., Choi, W. S., Lee, K. H., & Choi, D. K., (2010). Protective effects of *Gastrodia elata* Blume on MPPþ-induced cytotoxicity in human dopaminergic SH-SY5Y cells. *J. Ethnopharmacol.*, *130*(20), 290–298.
4. An, S. J., Park, S. K., Hwang, I. K., Choi, S. Y., Kim, S. K., Kwon, O. S., Jung, S. J., et al., (2003). Gastrodin decreases immunoreactivities of aminobutyric acid shunt enzymes in the hippocampus of seizure-sensitive gerbils. *J. Neurosci. Res.*, *71*, 534–543.
5. Anisuzzman, M., Zilani, M. N. H., Khushi, S. S., Asaduzzman, M., & Hossain, M. G., (2016). Antioxidant, Antibacterial potential and HPLC analysis of *Dioscorea alata* bulb. *Indonesian J. Pharm.*, *27*(1), 9–14.
6. Babu, P. N., Nagaraju, B., Yamini, K., Dhananjaneyulu, M., Venkateswarlu, K., & Mubina, M., (2014). Evaluation of antidepressant activity of ethanolic extract of *Dacus carota* in mice. *Journal of Pharmaceutical Sciences and Research*, *6*(2), 73.
7. Bahrami, R., Ghobadi, A., Behnoud, N., & Akhtari, E., (2018). Medicinal properties of *Daucus carota* in traditional Persian medicine and modern phytotherapy. *J. Biochem. Tech.*, *9*(2), 107–114.
8. Charmakar, S., Kunwar, R. M., Sharma, H. P., Rimal, B., Baral, S., Joshi, N., Gauli, K., et al., (2021). Production, distribution, use and trade of *Valeriana jatamansi* Jones in Nepal. *Global Ecology and Conservation*, *30*, e01792.
9. Chen, X., Krug, L., Yang, M., Berg, G., & Cernava, T., (2021). The Himalayan onion (*Allium wallichii* Kunth) harbors unique spatially organized bacterial communities. *Microbial. Ecology*, *82*, 909–918. doi: 10.1007/s00248-021-01728-5.
10. Chen, P. J., Hsieh, C. L., Su, K. P., Hou, Y. C., Chiang, H. M., & Sheen, L. Y., (2009). Rhizomes of *Gastrodia elata* B(L) possess antidepressant like effect via monoamine modulation in subchronic animal model. *Am. J. Chin. Med.*, *37*, 1113–1124.
11. Chen, P. J., & Sheen, L. Y., (2011). Gastrodiae rhizoma (天麻 tiānmá): A review of biological activity and antidepressant mechanisms*J. Tradit. Complement. Med.*, *1*(1), 31–40.
12. Chen, W. C., Lai, Y. S., Lin, S. H., Lu, K. H., Lin, Y. E., Panyod, S., Ho, C. T., & Sheen, L. Y., (2016). Anti-depressant effects of *Gastrodia elata* Blume and its compounds gastrodin and 4-hydroxybenzyl alcohol, via the monoaminergic system and neuronal cytoskeletal remodeling. *Journal of Ethnopharmacology*, *182*, 190–199.
13. Chen, Y. B., & Xue, Z., (1987). Study on chemical constituents of *Morinda officinalis* how. *Bull. Chin. Mat. Med.*, *12*, 27–29.
14. Chen, Z., Fan, D. N., & Ji, M. H., (2003). Effect of *Morinda officinalis* how. decoction on ilmnune function in mice. *Bull. Sci. Technol.*, *19*, 245–246.

15. Chhetri, H. B., & Gupta, V. N. P., (2007). A survey of non-timber forest products (NTFPS) in upper mustang. *Scientific World, 5*(5), 89–94. doi: 10.3126/sw.v5i5.2663.
16. Dhiman, B., Sharma, P., Shivani, & Pal, P. K., (2020). Biology, chemical diversity, agronomy, conservation, and industrial importance of *Valeriana jatamansi*: A natural sedative. *J. Appl. Res. Med. Aromat. Plants, 16*, Article 100243.
17. Dong, F. W., Jiang, H. H., Yang, L., Gong, Y., Zi, C. T., Yang, D., Ye, C. J., Li, H., Yang, J., Nian, Y., et al., (2018). Valepotriates from the roots and rhizomes of *Valeriana jatamansi* Jones as novel N-type calcium channel antagonists. *Front Pharmacol., 9*, 1–9.
18. Efferth, T., & Kaina, B., (2011). Toxicities by herbal medicines with emphasis to traditional Chinese medicine. *Current Drug Metabolism, 12*(10), 989–996.
19. Gaire, B. P., & Subedi, L., (2011). Medicinal plant diversity and their pharmacological aspects of Nepal Himalayas. *Pharmacognosy Journal, 3*(25), 6–17.
20. Hsieh, C. L., Chiang, S. Y., Cheng, K. S., Lin, Y. H., Tang, N. Y., Lee, C. J., Pon, C. Z., & Hsieh, C. T., (2001). Anticonvulsive and free radical scavenging activities of *Gastrodia elata* Bl. in kainic acid-treated rats. *Am. J. Chin. Med., 29*(20), 331–341.
21. Hsieh, C. L., Lin, J. J., Chiang, S. Y., Sa, S. Y., Tang, N. Y., Lin, G. G., Lin, I. H., et al., (2007). *Gastrodia elata* modulated activator protein 1 via c-Jun N-terminal kinase signaling pathway in kainic acid-induced epilepsy in rats. *J. Ethnopharmacol., 19*(2), 241–247.
22. Huang, D., Li, Q., Zhou, C., Zhou, S., & He, X., (2014). Intraspecific differentiation of *Allium wallichii* (*Amaryllidaceae*) inferred from chloroplast DNA and internal transcribed spacer fragments. *Journal of Systematics and Evolution, 52*, 341–354. doi: 10.1111/jse.12050.
23. Hwang, S. M., Lee, Y. J., Kang, D. G., & Lee, H. S., (2009). Anti-inflammatory effect of *Gastrodia elata* rhizome in human umbilical vein endothelial cells. *Am. J. Chin. Med., 37*(2), 395–406.
24. Jung, J. W., Yoon, B. H., Oh, H. R., Ahn, J. H., Kim, S. Y., Park, S. Y., & Ryu, J. H., (2006). Anxiolytic-like effects of *Gastrodia elata* and its phenolic constituents in mice. *Biological and Pharmaceutical Bulletin, 29*(2), 261–265.
25. Khan, A. U., & Gilani, A. H., (2011). Pharmacological basis for the medicinal use of *Valeriana wallichii* in constipation. *Lat Am J Pharm., 30*(1), 186–188.
26. Khan, M., Kumar, S., & Hamal, I. A., (2009). Medicinal plants of Sewa River catchment area in the Northwest Himalaya and its implication for conservation. *Ethnobot. Leafl., 13*, 1113–1139.
27. Kien, P. V., Son, D. C., Cuong, B. H., Ngoc, T. M., Thanh, T. G., Khanh, T. D., Tam, P. T. M., & Hung, T. V., (2020). Genetic diversity of *Morinda officinalis* Based on ITS and matK, rbcL and interspace trnH-psbA genes on the chloroplast. *Journal of Scientific Engineering and Research, 7*, 57–64.
28. Kim, H. J., Lee, S. R., & Moon, K. D., (2003). Ether fraction of methanol extracts of *Gastrodia elata*, medicinal herb protects against neuronal cell damage after transient global ischemia in gerbils. *Phytother. Res., 17*(8), 909–912.
29. Kim, I. T., Park, H. J., Nam, J. H., Park, Y. M., Won, J. H., Choi, J. W., Choe, B. K., & Lee, K. T., (2005). *In-vitro* and *in-vivo* anti-inflammatory and antinociceptive effects of the methanol extract of the roots of *Morinda officinalis*. *J. Pharm. Pharamcol., 57*, 607–615.

30. Kunwar, R. M., Nepal, B. K., Sigdel, K. P., & Balami, N., (2006). Contribution to the ethnobotany of the Dhading district, Nepal. *J. Sci. Technol., 7*, 65–69.
31. Leja, M., Kamińska, I., Kramer, M., Maksylewicz-Kaul, A., Kammerer, D., Carle, R., & Baranski, R., (2013). The content of phenolic compounds and radical scavenging activity varies with carrot origin and root color. *Plant Foods for Human Nutrition, 68*, 163–170.
32. Li, J., Peng, L., Huang, H. H., & Cao, H., (2009). Research progress of *in vitro* model building of blood–brain barrier. *Acad. J. Guangdong Coll. Pharm., 30*, 647–648.
33. Li, N., Wang, H. M., Guo, S. H., Lin, X., Zheng, L. P., & Wang, L., (2008b). Protection of apoptosis of osteobiast cultured *in vitro* by Morinda root polysaccharide. *China J. Orthop.-Trauma Jan., 21*, 39–41.
34. Li, X., Ma, S. Y., Cheng, D., Chang, H. S., Li, L. L., & Lu, Y., (2017). Research of *Morinda officinalis* how's oligosaccharide extraction and antidepressant effects. *Bulgar. Chem. Commun., 49*, 162–167.
35. Li, Y. F., Liu, Y. Q., Yang, M., Wang, H. L., Huang, W. C., Zhao, Y. M., & Luo, Z. P., (2004). The cytoprotective effect of inulin-type hexasaccharide extracted from *Morinda officinalis* on PC12 cells against the lesion induced by corticosterone. *Life Sci., 75*, 1531–1538.
36. Li, Y. F., Yuan, L., Xu, Y. K., Yang, M., Zhao, Y. M., & Luo, Z. P., (2001). Antistress effect of oligosaccharides extracted from *Morinda officinalis* in mice and rats. *Acta Phamacol. Sin., 22*, 1084–1088.
37. Lin, S., Shen, Y. H., Li, H. L., Yang, X. W., Chen, T., Lu, L. H., Huang, Z. S., Liu, R. H., Xu, X. K., Zhang, W. D., et al., (2009). Acylated iridoids with cytotoxicity from *Valeriana jatamansi. J Nat Prod., 72*(4), 650–655.
38. Lin, S. H., Chen, W. C., Lu, K. H., Chen, P. J., Hsieh, S. C., Pan, T. M., Chen, S. T., & Sheen, L. Y., (2014). Down-regulation of slit-robo pathway mediating neuronal cytoskeletal remodeling processes facilitates the antidepressive-like activity of *Gastrodia elata* Blume. *J. Agric. Food Chem., 62*(43), 10493–10503.
39. Liu, J., & Mori, A., (1992). Antioxidant and free radical scavenging activities of *Gastrodia elata* Bl. and *Uncaria rhynchophylla* (Miq.) jacks. *Neuropharmacology, 31*(12), 1287–1298.
40. Liu, J. J., (2011). Effect of polysaccharides from *Morinda offcinalis* on the oxidative stress and cognitive-behaviors in an experimental depression model of rats. *China Mod. Dr., 49*, 2–5.
41. Martins, J., & Brijesh, S., (2018). Phytochemistry and pharmacology of anti-depressant medicinal plants: A review. *Biomedicine & Pharmacotherapy, 104*, 343–365.
42. Mishra, M., Huang, J., Lee, Y. Y., Chua, D. S., Lin, X., Hu, J. M., & Heese, K., (2011). *Gastrodia elata* modulates amyloid precursor protein cleavage and cognitive functions in mice. *Biosci. Trends, 5*(3), 129–138.
43. Mohan, M., Haider, S. Z., Gautam, S. S., & Singh, P., (2019). *Allium stracheyi* (Baker), the endangered and traditional medicinally important herb of Uttarakhand Himalaya, India: A review. *Research in Pharmacy, 9*, 1–5. doi: 10.25081/rip.2019.v9.548.

44. Nighot, P. K., Leung, L., & Ma, T. Y., (2017). Chloride channel ClC- 2 enhances intestinal epithelial tight junction barrier function via regulation of caveolin-1 and caveolar trafficking of occludin. *Exp Cell Res.*, *352*(1), 113–122.
45. Nisha, P., Mishra, V. K., & Nema, N., (2021). Assessment of antidepressant and anxiolytic activities of aqueous extract of *Daucus carota* peel. *Asian Journal of Pharmacy and Pharmacology*, *7*(3), 143–149.
46. Ojemann, L. M., Nelson, W. L., Shin, D. S., Rowe, A. O., & Buchanan, R. A., (2006). Tian ma, an ancient Chinese herb, offers new options for the treatment of epilepsy and other conditions. *Epilepsy Behav.*, *8*(2), 376–383.
47. Pandey, N., Rijal, S., Adhikari, H., Bhantana, B., & Adhikhari, M., (2020). Production economics and determinants of carrot (*Daucus carota* L.) production in Chitwan, Nepal. *International Journal of Social Sciences and Management*, *7*(4), 234–241.
48. Park, H., Kim, S., Kim, S., Song, Y., Seung, K., Hong, D., Khang, G., & Lee, D., (2010). Antioxidant and anti-inflammatory activities of hydroxybenzyl alcohol releasing biodegradable polyoxalate nanoparticles. *Biomacromolecules, 11*(8), 2103–2108.
49. Peng, Z., Wang, S., Chen, G., Cai, M., Liu, R., Deng, J., Liu, J., et al., (2015). Gastrodin alleviates cerebral ischemic damage in mice by improving antioxidant and anti-inflammation activities and inhibiting apoptosis pathway. *Neurochem. Res., 40,* 661–673.
50. Penjor, D., Tshering, T., Bhattaria, G., & Namgay, T., (2020). The study of ethnobotanical uses by local healers in Taktse Chiwog from Central Bhutan. *Asian Plant Research Journal, 6*(4), 19–39.
51. Prasad, R., Naime, M., Routray, I., Mahmood, A., Khan, F., & Ali, S., (2010). *Valeriana jatamansi* partially reverses liver cirrhosis and tissue hyperproliferative response in rat. *Methods Find Exp Clin Pharmacol.*, *32*(10), 713–719.
52. Pyakurel, D., Sharma, I. B., & Ghimire, S. K., (2017). Trade and conservation of medicinal and aromatic plants in Western Nepal. *Bot. Orient.*, *11*, 27–37.
53. Raina, A. P., & Negi, K. S., (2015). Essential oil composition of *Valeriana jatamansi* Jones from Himalayan regions of India. *Indian Journal of Pharmaceutical Sciences*, *77*(2), 218.
54. Rana, M. S., & Samant, S. S., (2010). Threat categorization and conservation prioritization of floristic diversity in the Indian Himalayan region: A state of art approach from Manali wildlife sanctuary. *Journal for Nature Conservation, 18*, 159–168. doi: 10.1016/j.jnc.2009.08.004.
55. Reynolds, E. H., (2003). Brain and mind: A challenge for WHO. *The Lancet*, *361*(9373), 1924, 1925.
56. Sah, S. P., Mathela, C. S., & Chopra, K., (2010). Elucidation of possible mechanism of analgesic action of *Valeriana wallichii* DC chemotype (patchouli alcohol) in experimental animal models. *Indian J Exp Biol.*, *48*(3), 289–293.
57. Saini, N., Hedau, N. K., Khar, A., Yadav, S., Bhatt, J. C., & Agrawal, P. K., (2015). Successful deployment of marker assisted selection (MAS) for inbred and hybrid development in long-day onion (*Allium cepa* L.). *Indian Journal of Genetics and Plant Breeding*, *75*, 93–98.

58. Sakakibara, H., Yoshino, S., Kawai, Y., & Terao, J., (2008). Antidepressant-like effect of onion (*Allium cepa* L.) powder in a rat behavioral model of depression. *Bioscience, Biotechnology, and Biochemistry*, *72*(1), 94–100.
59. Samad, N., & Saleem, A., (2018). Administration of *Allium cepa* L. bulb attenuates stress-produced anxiety and depression and improves memory in male mice. *Metabolic Brain Disease*, *33*(1), 271–281.
60. Samant, S. S., Dhar, U., & Palni, L. M. S., (1998). *Medicinal Plants of Indian Himalaya: Diversity Distribution Potential Values.* G.B. Pant Institute of Himalayan Environment and Development; Almora, India.
61. Sayers, J., (2001). The world health report 2001-Mental health: New understanding, new hope. *Bulletin of the World Health Organization*, *79*, 1085–1085.
62. Sharma, N., Sharma, A. R., Patel, S. D., & Shrestha, K., (2016). Investigation on phytochemical, antimicrobial activity, and essential oil constituents of *Nardostachys jatamansi* DC. in different regions of Nepal. *J. Coast. Life Med.*, *4*(1), 56–60.
63. Shi, S. N., Shi, J. L., Liu, Y., Wang, Y. L., Wang, C. G., Hou, W. H., & Guo, J. Y., (2014). The anxiolytic effects of valtrate in rats involves changes of corticosterone levels. *Evidence-Based Complementary and Alternative Medicine, 2014.*
64. Singh, G., & Rawat, G. S., (2011). Ethnomedicinal survey of Kedarnath wildlife sanctuary in Western Himalaya, India. *Indian Journal of Fundamental and Applied Life Science, 1*(1), 35–46.
65. Sofowora, A., Ogunbodede, E., & Onayade, A., (2013). The role and place of medicinal plants in the strategies for disease prevention. *African Journal of Traditional, Complementary, and Alternative Medicines*, *10*(5), 210–229.
66. Subhan, F., Karim, N., Gilani, A. H., & Sewell, R. D. E., (2010). Terpenoid content of *Valeriana wallichii* extracts and antidepressant-like response profiles. *Phytother Res.*, *24*(5), 686–691.
67. Tang, Y., Liu, X., & Yu, B., (2002). Iridoids from the rhizomes and roots of *Valeriana jatamansi*. *J. Nat. Prod.*, *65*(12), 1949–1952.
68. Tiwari, U., Adams, S. J., Begum, N., Krishnamurthy, K. V., Ravikumar, K., & Padma, V., (2014). Pharmacognostic studies on two Himalayan species of traditional medicinal value: *Allium wallichii* and *Allium stracheyi*. *Notulae Scientia Biologicae*, *6*(2), 149–154. doi: 10.15835/nsb629308.
69. Tsai, H., Yan, J., Wang, S., & Chang, S., (2009). Neuroprotection against hyperhomocysteinemia-induced selective oxidative stress in brain regions of rats with folic acid or *Dioscorea alata* supplementation. *Curr. Top Nutraceutical Res.*, *7*(2), 73–80.
70. Wang, L., Sun, Y., Zhao, T., Li, Y., Zhao, X., Zhang, L., Wu, L., Zhang, L., Zhang, T., Wei, G., et al., (2020). Antidepressant effects and mechanisms of the total iridoids of *Valeriana jatamansi* on the brain-gut axis. *Planta Med.*, *86*(3), 172–179.
71. Wang, X., Yan, S., Wang, A., Li, Y., & Zhang, F., (2014). Gastrodin ameliorates memory deficits in 3,3′-iminodipropionitrile-induced rats: Possible involvement of dopaminergic system. *Neurochem. Res.*, *39*(8), 1458–1466.
72. Wang, Y. F., Li, Y. H., Xing, S. Q., Li, Y., Li, Z. T., Shang, X. D., Zhao, D. F., & Bai, L. Q., (2016). Review of experiment research progress in treating deficiency of kidney-yang syndrome by *Morinda officinalis* How. and its effective components. *CJTCMP, 31*, 5165–5167.

73. Xu, K. K., Lin, Y., Zhang, R., Lan, M., Chen, C., Li, S., Zuo, C., et al., (2015). Evaluation of safety of iridoids rich fraction from *Valeriana jatamansi* Jones: Acute and sub-chronic toxicity study in mice and rats. *J. Ethnopharmacol.*, *172*, 386–394.
74. Xu, L. Z., Xu, D. F., Han, Y., Liu, L. J., Sun, C. Y., Deng, J. H., Zhang, R. X., & Yuan, M., (2017). BDNF-GSK-3β-β-catenin pathway in the mPFC is involved in antidepressant-like effects of *Morinda officinalis* oligosaccharides in rats. *Int. J. Neuropsychopharmacol.*, *20*, 83–93.
75. Yan, Z., Zhang, T., Xiao, T., et al., (2010). Anti-anxiety effect of *Valeriana jatamansi* Jones extract via regulation of the hypothalamus-pituitary-adrenal axis. *Neural Regeneration Research, 5*, 1071–1075.
76. Yan, Z. Y., Zhang, T. E., Pan, L., et al., (2011). Action of *Valeriana jatamansi* Jones on the apoptosis-related genes expression in the anxiety model of rat. *Procedia Environ. Sci., 8*, 744–750.
77. Yang, Y., Kimura-Ohba, S., Thompson, J. F., Salayandia, V. M., Cosse, M., Raz, L., Jalal, F. Y., & Rosenberg, G. A., (2018). Vascular tight junction disruption and angiogenesis in spontaneously hypertensive rat with neuroinflammatory white matter injury. *Neurobiol Dis., 114*, 95–110.
78. Yang, X., Guo, J. Y., Jiang, Y. N., Liu, M. M., Li, Q. Y., Li, J. Y., & Shi, J. L., (2021). *Valeriana jatamansi* Jones ex roxb. against post-traumatic stress disorder, network pharmacological analysis, and *in vivo* evaluation. *Frontiers in Pharmacology, 12.*
79. Ye, W. H., Zou, Z. J., & Gong, M. J., (2013). Metabonomic study of anti-inflammatory effect of *Morinda officinalis* how. on acute inflammation induced by carrageenan. *Pharm. Clin. Chin. Mater. Med., 4*, 22–26.
80. Yu, S. S., Zhao, J., Zheng, W. P., & Zhao, Y., (2010). Neuroprotective effect of 4-hydroxybenzyl alcohol against transient focal cerebral ischemia via anti-apoptosis in rats. *Brain Res., 1308*, 167–175.
81. Zeng, X., Zhang, S., Zhang, L., Zhang, K., & Zheng, X., (2006). A study of the neuroprotective effect of the phenolic glucoside gastrodin during cerebral ischemia *in vivo* and *in vitro*. *Planta Med., 72*(15), 1359–1365.
82. Zhai, X., Kong, Z., Wang, S., et al., (2016). Study on anti-anxiety activity of extract and total valepotriate in Valerianae Rhizoma. *Chin. Tradit. Herb. Drugs, 47*, 1361–1365.
83. Zhan, H. D., Zhou, H. Y., Sui, Y. P., Du, X. L., Wang, W. H., Dai, L., Sui, F., Huo, H. R., & Jiang, T. L., (2016). The rhizome of *Gastrodia elata* Blume–an ethnopharmacological review. *Journal of Ethnopharmacology, 189*, 361–385.
84. Zhang, H., Li, J., Xia, J., & Lin, S., (2013). Antioxidant activity and physicochemical properties of an acidic polysaccharide from *Morinda officinalis. Int. J. Biol. Macromol., 58*, 7–12.
85. Zhang, H. L., Li, J., Li, G., Wang, D. M., Zhu, L. P., & Yang, D. P., (2009). Structural characterization and anti-fatigue activity of polysaccharides from the roots of *Morinda officinalis. Int. J. Biol. Macromol., 44*, 257–261.
86. Zhang, J. H., Xin, H. L., Xu, Y. M., Shen, Y., He, Y. Q., Lin, B., & Du, J., (2018). *Morinda officinalis*—A comprehensive review of traditional uses, phytochemistry and pharmacology. *Journal of Ethnopharmacology, 213*, 230–255.

87. Zhang, L., Wang, L., Huang, L., Zhao, Y., Ding, H., Li, B., Wen, L., et al., (2021). Antidepressant effects of total iridoids of *Valeriana jatamansi* via the intestinal flora-blood–brain barrier pathway. *Pharmaceutical Biology*, *59*(1), 910–919.
88. Zhang, Y. Z., Li, Y. F., & Liu, G., (2005). Antidepressant effect of oligosaccharides extracted from *Morinda officinalis* on the learned helplessness rat model. *Chin. J. Behav. Med. Sci.*, *14*, 309–311.
89. Zhang, Z. Q., Huang, S. J., Yuan, L., Zhao, N., Xu, Y. K., Yang, M., Luo, Z. P., et al., (2001). Effects of *Morinda oficinalis* oligosaeeharides on performance of the swimming tests in mice and rats and the learned helplessness paradigm in rats. *Chin. J. Pharmacol. Toxic.*, *15*, 262–265.
90. Zhang, Z. Q., Yuan, L., Zhao, N., Xu, Y. K., Yang, M., & Luo, Z. P., (2000a). Antidepressant effect of the ethanolic extracts of the roots of *Morinda officinalis* in rats and mice. *Chin. Pharm. J.*, *35*, 739–741.
91. Zou, L. Y., & Zhang, H. Y., (2012). Research advance of *Morinda officinalis* oligosaccharidesin treatment of depression. *Chin. J. New Drugs*, *21*, 1889–1891.

## CHAPTER 5

# Antidepression Agents in Himalayan Grains, Beans, and Nuts

JIGISHA ANAND,[1] KIRAN DHAMI,[2] PRABHAKAR SEMWAL,[2] and NISHANT RAI1

[1]*Department of Biotechnology, Graphic Era Deemed to be University, Dehradun, Uttarakhand, India*

[2]*Department of Life Sciences, Graphic Era Deemed to be University, Dehradun, Uttarakhand, India*

## ABSTRACT

The state of our mind and mental processes is influenced by individual well-being. Depression is a mood disorder characterized by dysfunctional social behavior, which has a profound effect on a person's social performance, productivity, and relationships with family and friends. The etiology of depression includes several social, psychological, and biological factors that elicit a state of recurrent or persistent melancholy, dramatically affecting a person's ability to function and live a spirited and vivacious life. Several psychological and pharmacological treatments for depression exist, pertaining to acute phase treatment and continuation phase treatment. However, substantial side effects such as headaches, sleep problems, nausea, dizziness, restlessness, trembling, sexual dysfunction, and addiction may be associated with these conventional medications. As evident from the research literature, the majority of natural herbal phyto-compounds have been experimentally explored for their promising role as antidepressants and their helpfulness in the management of depression and its associated mental disorders. These herbal antidepressants

Himalayan Medicinal Plants for the Treatment of Depression: A Source of Rich Antidepressant Agents.
Abdur Rauf, Prabhakar Semwal, & Hafiz Ansar Rasul Suleria (Eds.)

are perceived as comparatively safer alternatives and have a lower risk of withdrawal or side effects. This chapter aims to review the literature evidence highlighting the potential of Himalayan grains, beans, and nuts with nutraceutical and medicinal significance for improving depression and associated mental disorders. We hope that this collective information would be clinically relevant to researchers and may be worthy of future investigation on elucidating the mechanism of action, molecular interactions, optimal dosage, and safety of Himalayan botanical resources for efficient use as antidepressant drugs.

## 5.1 INTRODUCTION

Depression is now becoming a common and serious medical illness, which is a mental disorder that negatively affects one's feelings, thinking, and actions. It is considered a disability and a fatality, harming humanity, relationships, and personalities globally [78]. Disability results from the interaction between individuals with a health condition, including depression, which is now common in over 1 billion people [112].

Depression may include symptoms such as somatic symptoms and cognitive symptoms, such as changes in appetite, alterations in weight, feelings of death or suicide, and a negative view of oneself, among others. Depression can occur at any time during the lifespan. Depression, along with anxiety, is one of the most fundamental emotional states of a conscious being. According to the World Health report, almost 350 million people globally are affected by depression, making it the second-largest disease after heart disease [112].

As a chronic and incapacitating mental condition, depression affects 15–20% of the world's population, making a substantial contribution to the overall burden of sickness, with 3.8% of the affected human population, including 5% of adults, as per the WHO report [38, 44]. It is not possible to pinpoint a single component that causes or prevents depression in all cases [14]. As a result, the development of effective antidepressants will have significant social, economic, and health implications. Therefore, a better therapeutic approach could be followed to identify each patient's unique reasons and then apply precise actions, not only to alleviate sadness but also to address the body's dysfunctions that cause depressive symptoms [116].

This book chapter summarizes the literature evidence highlighting the nutraceutical and therapeutic significance of Himalayan grains, beans, and nuts in curing depression and associated mental disorders.

## 5.2 DEPRESSION: A STATE OF MIND AND BODY

Mental health problems remain key contributors to the global burden of disease (GBD) [112]. The COVID-19 epidemic has had an unprecedented impact on the mental and physical health of healthcare workers, patients, and related individuals. Mental health concerns such as anger, fatigue, depression (4–31%), post-traumatic stress disorder (PTSD, 12.1–46.9%), and anxiety (6.5–63%) have taken a significant toll on people's mental well-being. According to data from the WHO, depression and anxiety have increased by 25% since the beginning of the pandemic, affecting an additional 1 billion people with mental disorders. Consequently, COVID-19 has highlighted the gaps in the capacity of healthcare systems worldwide to address mental health conditions.

According to a study, India has over 40 million people experiencing depression due to various reasons. Furthermore, lack of awareness and the stigma associated with mental health issues exacerbate the profound personal and societal impact of depression [82]. The mind-body connection has always intrigued medical professionals, as the ultimate measure of successful treatment lies in the well-being of the mind rather than the body. Common mental disorders (CMD), including depression and anxiety, are considered part of the spectrum of non-communicable diseases (NCDs). Comorbidity of anxiety and depression with hypertension, diabetes, and cardiovascular disease is prevalent in India [18]. Additionally, numerous studies have shown that serious/critical medical treatments also have a negative impact on patients, with approximately one-third of patients undergoing cardiac surgery experiencing depression at the time of discharge.

Depression is prevalent in preoperatively dejected patients who tend to have perfect depression scores, whereas other patients have deteriorated depressive symptoms. Major depressive disorder (MDD) is classically deliberated as a mental illness. Additionally, MDD is associated with a high frequency of emerging serious medical comorbidities such as heart diseases or stroke, diabetes, osteoporosis, dementia, and metabolic syndromes. These conditions generally occur later in life, and it has been proposed that "accelerated aging" is correlated to MDD [93]. Depression is linked to changes in brain structure and functions. The brain's capacity to alter its structure and function is referred to as brain plasticity, which is also the physiological source for the psychological and behavioral version. The study of variations in brain plasticity of a depressed patient is now easier with the advances in medical imaging systems.

More studies have now been conducted at the molecular, cellular, behavioral, and structural network levels. It has been investigated that abnormalities related to structure and functions have appeared in key brain regions of depression patients, which also include volume changes and functional damage [117].

In various regions of India, the occurrence of postpartum depression (PPD) ranges from 15% to 20%. PPD typically begins two weeks to a month after delivery and can persist for months or even up to a year. According to the WHO report, depression affects 10% of pregnant women globally and 13% of women in the immediate postpartum period.

Currently, there are numerous methods to treat depression, but drug therapy is the primary approach. However, it comes with various side effects, the risk of addiction, high cost, and poor patient compliance. Drug treatment is considered unsatisfactory as it significantly impacts the quality of the patient's life [116].

Yoga, originating in ancient India, and exercise are recognized as alternative medicine practices that incorporate mind-body techniques [96]. Additionally, in ancient times, natural foods found in the Himalayan regions were commonly used as antidepressants, and people possessed traditional knowledge about them. Therefore, this study aims to highlight the various aspects of Himalayan foods as potential antidepressants.

## 5.3 TRADITIONAL KNOWLEDGE OF FOODS WITH ANTIDEPRESSANT ACTIVITY

Depressive disorder has been recognized as a mental illness since ancient times. Hippocrates, the father of medicine, characterized its clinical picture and established the term "melancholia," still used in present times. This common, serious, and deadly illness was virtually untreatable until the middle of the $20^{th}$ century [69]. Since the late 1950s, antidepressant drugs have been available to mankind. Iproniazid (a monoamine oxidase inhibitor), which was originally developed as an anti-tuberculous agent [13], and imipramine, which was first developed as an antipsychotic drug in 1958, were the first effective antidepressant drugs.

Pharmacotherapy and psychotherapy are the most common treatments for depression and anxiety-related disorders. However, antidepressants act as modulators of neurotransmitters that play an important role in the functioning of both the central nervous system (CNS) and peripheral nervous system (PNS). The current medications are not very effective for cancer patients [67] and have shown substantial side effects such as seizures, sexual dysfunction, headaches, etc. [32].

Our nature has blessed human beings with an enormous treasure of natural products that have proven to be vital components of mankind. These include various fruits, nuts, cereals, vegetables, leaves, seeds, etc. These natural products contain diverse phytoconstituents that naturally have multiple benefits for human health, including normal brain growth and functioning. According to the World Health Organization (WHO), there is always exploration of natural alternative medicine as Natural Products. Thus, plants/herbs not only fulfill the need for food, but now these natural products are recognized for their multiple beneficial health effects, medicinal values, and nutraceutical properties, especially in the present age of evolving drug resistance and the indescribable side effects of chemical drugs [6, 25].

Owing to the plant diversity in India and their multiple benefits as food and medicine, India has become a hub for medicinal plants and natural medicine. Many modern medicines are derived from plants. The primary healthcare needs of people in remote areas of India are largely met through the consumption of traditional medicine such as Ayurveda, Siddha, and Unani, in various formulations [84].

Herbal remedies and plant products as therapeutic agents have gained popularity in industrialized societies for the management of depression and anxiety [63]. Recently, the commercialization of natural foods with antidepressant properties has become increasingly established in the global health sector. Numerous research contributions in ethnopharmacology have generated and updated data related to natural products, including antidepressants [21].

Some of the major explored plants having antidepressant activity are enlisted in Table 5.1.

**TABLE 5.1** Traditional and Complementary Medicines Against Depression

| SL. No. | Name of Species | Plant Part Used in Medicine | Developed Drug | Different Classes of Bioactive Compounds | References |
|---|---|---|---|---|---|
| 1. | *Hypericum perforatum* L. | Flowers and dried upper part of the plant. | *Hperichiherba* | Anthraquinone, phloroglucinol, flavonoids, phenolic compounds, xanthones, and tannins. | [40] |
| 2. | *Valeriana officinalis* L. | Rhizomes, roots, and stolons. | *Valerianae radix* | Valepotriates, resin, lignans, alkaloids, tannins, flavonoids, phenolic acids, essential oil. | [40] |
| 3. | *Ocimum basilicum* L. | Leaves, stem, and flowers. | *Basiliciherba* | Flavonoids, phenolic acids, phytosterols, essential oil. | [40] |
| 4. | *Melissa officinalis* L. | Leaves | *Lemon balm* | Hydroxycinnamic acids, flavonoids, pectin, tannins, triterpenes, resins, essential oil. | [40] |
| 5. | *Verbena officinalis* L. | Dried parts of stem, leaves, and flowers. | *Verbenaeherba* | Iridoid glycosides, triterpenes, flavonoids, phenylpropanoids. | [40] |
| 6. | *Eleutherococcus senticosus* (Rupr et Maxim.) Maxim | Roots and aerial parts of the plant. | – | Triterpenoid saponins, lignans, coumarins, flavonoids, phenolics, essential oil. | [41] |
| 7. | *Leonurus cardiaca* L. | Flowers and leaves | – | Terpenes, tannins, sterols, flavonoids, and phenolic acids. | [59] |
| 8. | *Valeriana edulis* Nutt. ex Torr. | Root and rhizomes | – | Iridoid valepotriates, alkaloids, lignans, essential oils. | [20] |
| 9. | *Galphimia glauca* Cav. | Leaves, flowers, stem. | – | Nor-seco-triterpenes. | [20, 29] |
| 10. | *Annona muricata* L. | Leaves and soursop | – | Alkaloids, acetogenins cyclohexapeptide, cyclopeptide. | [29] |
| 11. | *Salvadora persica* L. | Roots | – | Alkaloids, flavonoids, saponins, tannins. | [52] |
| 12. | *Perilla frutescens* | Essential oil, leaves | – | Monoterpene perillaldehyde (PAE) | [31] |

**TABLE 5.1** *(Continued)*

| SL. No. | Name of Species | Plant Part Used in Medicine | Developed Drug | Different Classes of Bioactive Compounds | References |
|---|---|---|---|---|---|
| 13. | *Lavandula angustifolia* | Essential oils/flower | – | Linalool, linalyl acetate, monoterpene alcohol lavandulol. | [63] |
| 14. | *Melissa officinalis* | Leaves, essential oil | – | Phenolics and flavonoids. | [105] |
| 15. | *Camellia sinensis* | Flavonoids, alkaloids | – | L-theanine, catechin, tea pigment and GABA. | [95] |
| 16. | *Boerhavia diffusa* Linn | Alkaloids | – | Punarnavine | [34] |
| 17. | *Foeniculum vulgare* | Seeds oil contains | – | α-phellandrene estragole, fenchone, and tansanethole, camphene, cymene, α-pinene, β-pinene, myrcene, γ-terpinene and panisaldehyde. | [1] |
| 18. | *Vanillus planifolia* | Beans | – | Vanillin (hydroxybenzaldehyde) | [4, 99] |
| 19. | *Garcinia indica* | Fruit, rind, seed | – | Garcinol, hydroxycitric acid (HCA), cyanidin-3-sambubioside, and cyanidin-3-glucoside. | [26, 55] |
| 20. | *Crocus sativus* L. | Stigma | – | Crocin, crocetin, safranal | [86, 101] |
| 21. | *Centella asiatica* (L.) | Leaves | – | Asiaticosides | [35, 100] |
| 22. | *Asparagus racemosus* | Roots | – | Saponins | [3, 100] |
| 23. | *Dolichos biflorus* | Seeds, leaves, stem | – | Streptogenin, β-sitosterol, bulbiformin, linoleic acid, polyphenols, oxalates. | [51, 71] |
| 24. | *Peganum harmala* L | Seeds, bark, root | – | Harmine, harmaline | [71, 106] |
| 25. | *Brassica juncea* L. | Leaves | – | Flavonoids, saponins, sterols/triterpenoids, and terpenes. | [58, 107] |
| 26. | *Nardostachys jatamansi* DC | Roots | – | 7-oxonardinoperoxide | [54, 90] |
| 27. | *Salvia hispanica* | Leaves, seed | – | Salvinorin A, Salvianolic acid B | [8, 56] |

## 5.4 ROLE OF HIMALAYAN FOODS AS ANTIDEPRESSANTNUTS

1. *Juglans regia:* Or Walnuts are one of the world's most venerable fruits, comprising more than 20 species. They are native to North and South America, Asia, southern Europe, and the West Indies, and are cultivated broadly by several ethnic groups in the Himalayan region. Walnuts consist of a plethora of conceivably neuroprotective substances such as folic acid, vitamin E, melatonin, several antioxidative polyphenolic compounds, and significant amounts of fatty acids. Therapeutic studies have indicated that the consumption of walnuts lowers depressive symptoms and can nourish brain cells to improve human memory. Melatonin, an active constituent in walnuts, promotes better sleep and relaxes the mind [81].

   A recent study validates that regular consumption of walnuts may reduce the spread and frequency of stress. Nuts have the edge when it comes to reducing symptoms of depression because they are also one of the richest sources of omega-3 fatty acids, which support overall brain health. A comparative study has shown that stress levels were 26% lower for walnut eaters compared to people who did not eat any kind of nuts [5].

   Several species of Juglans are used as long-term raw drugs. More recently, researchers have explored its effectiveness as an antidepressant. In human and animal model studies, diet supplements containing walnuts have shown improved cognitive power and reported reduced risk and/or progression of Mild Cognitive Impairment, Parkinson's disease, stroke, and Alzheimer's [19, 71, 110]. Scientifically, studies have further proven the beneficial effect of walnuts against cardiovascular disease and type 2 diabetes [49, 70, 109]. Additionally, a diet rich in walnuts has been recognized to possess antibacterial and anticancer effects [22] (Figure 5.1).

2. *Myrica esculenta:* It is popularly called Himalayan Bayberry; box berry or Kuphal is an important traditional plant in the sub-tropical Himalayan region. They are edible species, with every part being used as medicine to treat various ailments. They have also been a valuable economic resource for local communities [10]. The fruits, bark, leaves, and roots of these species contain high levels of vitamins, phenolics, flavonoids, terpenoids, glycosides, and volatile

**FIGURE 5.1** Fruit of *Juglans regia.*

*Source*: Photo by Fir0002 at English Wikipedia. https://creativecommons.org/licenses/by-sa/3.0/deed.en

oils. These natural antioxidants play a crucial role in reducing oxidative stress and preventing degenerative diseases [83, 103].

The roots of *M. esculenta* are used in the preparation of ayurvedic formulations such as Chwayanparsh and Brahmarasayan, which are known to enhance digestion, memory, and physical strength. Phytochemical studies have reported anti-inflammatory, antimicrobial, antidiabetic, antiallergic, and antioxidant properties of the leaf and root solvent extracts [47, 85, 91]. In a recent study, the methanolic extract of *M. esculenta* has shown significant improvement in muscle coordination and behavioral activity in haloperidol induced Wistar albino rat. The study revealed ameliorated levels of antioxidant enzymes and a reduction in glutathione levels **[46].**

The bark of the tree is regionally used as a disinfectant, which cleanses rotten wounds. Several studies have shown its important chemoprotective and antioxidant properties, envisioning its neuroprotective potential [11]. Flower oil is used as a tonic for inflammation and paralysis [53,103] (Figure 5.2).

3. *Ziziphus mauritiana:* It is popularly known as Indian jujube or Ber, cultivated widely in tropical countries like India. The Ber fruit is imbued with vitamins B and C, terpenoids, saponins, alkaloids, and fatty acids, and is loaded with high dietary fiber, vitamins, and minerals. Plenty of active phytoconstituents like kaempferol, quercetin, lanosterol, berberine, stigmasterol, and so forth, were revealed in the different analyses [2, 48, 80, 92, 113].

**FIGURE 5.2** Fruit of *Myrica esculenta.*

*Source*: https://creativecommons.org/licenses/by-sa/3.0/

The traditional use of Ziziphus leaves, stems, and fruits in respiratory, hair, and skin problems has been widely known. In addition, investigations have identified betulinic acid, triterpenic acid, flavonoids, and polysaccharides as their essential active phytocompounds, contributing to immune stimulation and hematopoietic functions.

The plant is widely recognized in traditional medicine as a natural remedy for the symptoms of depression and anxiety. In previous studies, the extracts of seeds and leaves of jujube have shown their ameliorating role in memory enhancement, improving learning capacity, and antidepressant activities were reported in recent studies [54, 115, 116]. In addition, the soothing properties of jujube fruit are also known to moderate cortisol hormones and help keep the mind and body relaxed (Figure 5.3).

4. *Eleusine coracana:* Or finger millet is the major crop of Himalayan regions and is considered one of the most nutritious cereals [7, 90]. Being the richest source of iron, calcium, and magnesium among all millets, ragi contains water and fat-soluble vitamins, carbohydrates comprising free sugars, starch, non-starchy polysaccharides, and high fiber [24, 37].

   Ragi is a free radical scavenger that acts as a natural relaxant. Previous research has shown that regular consumption of ragi

**FIGURE 5.3** *Ziziphus mauritiana* plant.

*Source*: Photo by Ethel Aardvark. https://creativecommons.org/licenses/by/3.0/deed.en

aids in dealing with conditions like anxiety, sleeping disorders, and depression [24]. The amino acid tryptophan present in ragi works as a natural relaxant and helps in controlling migraine attacks. The high tryptophan level is responsible for enhancing the condition of the nerve impulse by activating the memory center in the brain and relaxing the mind. In addition, studies have demonstrated the beneficial effect of finger millets against anxiety and insomnia by maintaining good moods and promoting sound sleep [98] (Figure 5.4).

**FIGURE 5.4** *Eleusine coracana* grains.

*Source*: Photo by Agnieszka Kwiecień, Nova. https://creativecommons.org/licenses/by-sa/4.0/deed.en

5. *Linum usitatissiumum:* Or flax is an active plant-based diet that provides healthy fats, antioxidants, and fiber. The pharmacological accomplishments of flax are mainly attributed to its rich composition, comprising phenols, sterols, fatty acids, phytoestrogenic lignans, omega-3 fatty acids, and antioxidants. These purified phytocompounds and different solvent extracts have been studied for their antimicrobial, anticancer, antidiabetic, anti-inflammatory, hepatoprotective, cognitive effects, and many more. Diverse in its function, flax has played a significant role in Ayurvedic medicine for 1,000 years. The phytonutrients in flax enhance head health by reducing free radical damage. It helps to increase blood circulation through the scalp, promoting hair growth and reducing hair loss [36].

   In clinical trials, the intervention of flaxseed oil in stress and depression management has been proposed. Enriched with Omega 3 fatty acids, flaxseed helps uplift and control overall brain health. Animal model-based studies have discovered that expanding the omega-3 fatty acids in the daily diet can help avoid major depression disorders [30]. Additionally, the studies have identified that flax oil supplementation led to a better mood ranking in females with depression disorder and could increase the concentration of brain-derived neurotrophic factor (BDNF), which is found to play an essential role in mood and cognitive function [75].

   Flaxseed oil is rich in tryptophan amino acids and omega-3 fatty acids, which improve serotonin production that may lead to the betterment of brain development and spatial memory [36]. Dietary flaxseed may also aid in normal brain function during situations of neural disease (Figure 5.5).

6. *Hypericum perforatum:* It is the only natural antidepressant. Famously known as St. John's wort, hypericum, or millepertuis, it is an herbaceous plant native to Western Asia, North Africa, and Europe. It possesses multiple external and internal medicinal effects. Studies have shown the relieving effect of its essential oils in wounds, burns, inflammation, etc., while internally it serves as a natural antidepressant that is effective in the treatment of several ailments, such as mildew, anxiety, and mild to moderate depression [7, 16].

   Several phytochemical analyses have revealed the role of major phyto-compounds naphthodianthrones, quercetin, rutin, hypericin,

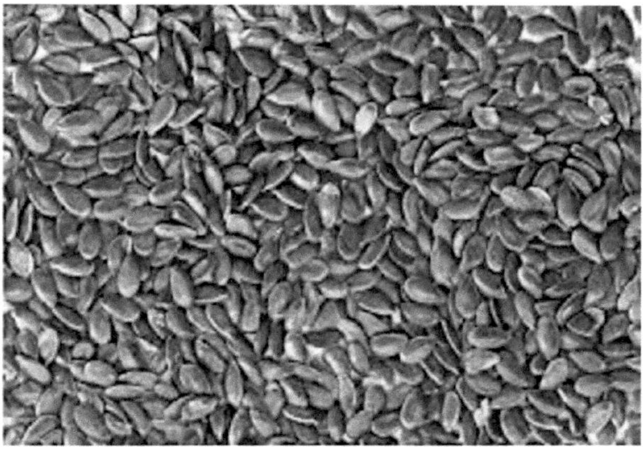

**FIGURE 5.5** *Linum usitatissiumum* seeds.

kaempferol, and pseudohypericin as neuroprotective [33, 66, 88, 114]. These bioactive compounds work in multiplicity rather than as an alone constituent or with a single mechanism of action [16, 28]. Their chemoprotective and antioxidant activity depicted reduced β-amyloid deposition, protection from motor deficit, improvement in motor coordination, and spatial learning acquisition [41, 102] (Figure 5.6).

**FIGURE 5.6** *Hypericum perforatum* plant.

*Source*: Photo by Agnieszka Kwiecień, Nova. https://creativecommons.org/licenses/by-sa/4.0/deed.en

7. *Zanthoxylum armatum:* It is one of the most important medicinal plants in India. About 10 varieties are cultivated in India. Almost every part of this plant is used in traditional Indian medicine to treat various ailments. It is locally called Tejphal (in Hindi) or Timur in Nepal and has ethnomedicinal significance [43]. Since antiquity, the bark, seed, and fruit of *Z. armatum* are used to treat fever, cough, cold, diarrhea, vertigo, and dysentery [9]. The pharmaceutically active phytocompounds like berberine, xanthoplanine, dictamnine, asarinin, fargesin, armatamid, α- and β-amyrins have been studied and identified to possess larvicidal, antioxidant, and anti-inflammatory effects.

   The neuroprotective activity of *Z. armatum* has been investigated in animal models treated with hexane extract. The test exhibited an antidepressant activity of *Z. armatum* extract conferred by its anti-inflammatory potential together with altering monoaminergic neurotransmitters in the brain [17] (Figure 5.7).

**FIGURE 5.7** *Zanthoxylum armatum* plants.

*Source*: Photo by Krzysztof Golik. https://creativecommons.org/licenses/by-sa/4.0/deed.en

8. *Phaseolus vulgaris:* It is one of the safe and healthy foods that accomplish us with an abundant source of protein, fibers, essential vitamins, minerals, and polyunsaturated fatty acids (PUFA) **[27, 104]**. Commonly referred to as black beans, they are highly favored legumes in both low-income and developed countries. Previous studies on the health benefits of black beans have highlighted their ability to enhance bone health, lower blood pressure, regulate diabetes, reduce the risk of

heart disease, prevent cancer, promote a healthy digestive system, aid in weight loss, and improve fertility [39, 65].

The literature also reports on some of the most significant biological functions associated with the chemical properties of beans. These include antioxidant properties, which make them beneficial for cardiovascular and neuronal protection, as well as attributes such as antimicrobial, antihyperglycemic, and anti-cancer activity [57, 104].

Notably, black beans are the source of tryptophan, which has been shown in several clinical and pre-clinical studies to improve the concentration of 5-hydroxytryptophan (5-HTP). 5-HTP is a synaptic monoamine and an important factor in controlling the central nervous system (CNS). In mice orally fed with an aqueous extract of *P. vulgaris,* the pharmacological treatment of depression has been shown to reduce immobility in the force swim test and potentiate 5-HTP [23]. However, there is not much literature available to validate the effectiveness of these nutritionally rich legumes in depression and other related mental illnesses (Figure 5.8).

**FIGURE 5.8** *Phaseolus vulgaris* seeds.

*Source*: Photo by Sanjay Acharya. https://creativecommons.org/licenses/by-sa/3.0/deed.en

9. *Mucuna pruriens:* It is a tropical and traditional Indian herb, popularly called velvet bean or dopamine bean. Its medicinal properties are well recognized in the Ayurvedic medical system for managing a number of diseases like Parkinson's and male infertility [83].

*M. pruriens* seeds are well known for their dopaminergic action and have several therapeutic applications in traditional medicine to treat or control nervous disorders. The high levels of L-DOPA in *M. pruriens* seeds are the essential source of dopamine [76]. Traditionally, the seeds are useful in dysmenorrhea, delirium, tuberculosis, diabetes, gout, sexual dysfunction, cough, weakness, rheumatic disorders, muscle pain, gonorrhea, infertility, and cancer [111]. The seeds of *M. pruriens* possess dopaminergic action and are responsible for the restoration of endogenous monoamine content, including dopamine, to the substantia nigra, indicating their neuroprotective activity [160].

The psycho-pharmacological study of seed powder of *M. pruriens* in depression mice and rat models deciphered potentiation of 5-Hydroxytryptophan-induced head twitches response and reversal of reserpine-induced hypothermia (RIH) [73]. The studies have also reported an increase in 5-HTP or serotonin concentrations in the synapse, indicating the antidepressant-like potential of *M. pruriens* [72] (Figure 5.9).

**FIGURE 5.9** *Mucuna pruriens* plant.

10. *Berberis aristate:* It is an herbal remedy useful in the medicinal system for the treatment of the CNS, depression, sadness, grief, stress, and tension [50, 68]. *B. aristata* is reported to have several other pharmacological effects in the case of stomach and amoebic dysentery, and possesses antimicrobial, antioxidant, anti-tumor, and anti-hyperglycemic properties [110–112].

    Berberine is an alkaloid isolated from *B. aristata* and has been identified as a key molecule. Recent reports have highlighted its role in various disorders related to the nervous system. A study has demonstrated the role of berberine in enhancing the anti-inflammatory effect of antidepressants like tranylcypromine, imipramine, fluoxetine, and norepinephrine (NE) [108].

    In another study, the researcher demonstrated the antidepressant effect of berberine chloride in different behavioral paradigms of despair, as studied in the Forced Swimming Test (FST) and tail suspension test (TST). The anti-immobility effect of berberine elucidated its antidepressant effect [50, 89]. Furthermore, Kulkarni & Dhir [50] reported the involvement of the nitric oxide pathway and sigma receptors in mediating its antidepressant-like activity. However, further extensive pharmacological study is required to understand the antidepressant activity of *B. arista* (Figure 5.10) and explore the possibilities of the role of other active ingredients and the mode of action.

**FIGURE 5.10** *Berberis aristate* fruit.

*Source*: Photo by Buddhika.jm. https://creativecommons.org/licenses/by-sa/3.0/

11. *Macrotyloma uniflorum:* Or horse gram is one of the lesser-known legumes of the Indian subcontinent and is commonly used as cattle feed. The legume is a rich source of iron and molybdenum and is known for its use in the treatment of renal stones, edema, and piles in Ayurveda [64].

    *M. uniflorum* is recommended for the treatment of depression in traditional Ayurveda, Siddha, and Unani medicine. The seeds of the plant *Macrotyloma uniflorum* have historically been used to treat many diseases, including neurodegenerative diseases. However, its role as a potential source in treating depression has not yet been scientifically proven. The seeds of Horsegram are extracted and available as an herbal remedy for depression (Figure 5.11).

**FIGURE 5.11** *Macrotyloma uniflorum* seeds.

*Source*: Photo by Sanjay Acharya. https://creativecommons.org/licenses/by-sa/3.0/deed.en

12. *Valeriana jatamansi:* It is cultivated in the temperate Himalayan region at altitudes ranging from 1,000 to 3,000 meters above sea level. Its active phytoconstituents include alkaloids, flavanones, hesperidin, Gamma-aminobutyric acid, sesquiterpenes, etc. [61, 94]. Historically, the roots and rhizome of *V. jatamansi* have been used medicinally to treat diarrhea, indigestion, high blood pressure, relieve pain, and act as a tranquilizing agent.

Regarding the healthy central nervous system (CNS), the gut microflora has been investigated and scientifically proven to regulate the integrity of the blood-brain barrier (BBB) by upregulating the expression of zonula occludens-1 (ZO-1) and claudin-5 [115]. The BBB is a selectively permeable barrier between the brain and brain tissues. Interestingly, studies have indicated that the progression of neurodegenerative diseases such as encephalitis, Alzheimer's, Parkinson's, autism, depression, and schizophrenia may be a result of damage caused to the BBB in response to changes in the natural microflora of the intestines [87, 97].

A recent study states that changes in gastrointestinal microbiome diversity may lead to the production of certain neurotransmitters and hormones, which could stimulate the enteric nervous system. More recently, the antidepressant effect of *V. jatamansi* via the BBB-intestinal flora pathway has been successfully demonstrated. The study successfully showed the action of total iridoids (TIV) of *V. jatamansi* in regulating the BBB pathway and intestinal microflora in depressed mouse models and provided the possible basis for its antidepressant effect [108, 116]. The study reported that *V. jatamansi* administration maintains the balance of dominant bacterial species in the treated mice (Figure 5.12).

**FIGURE 5.12** *Valeriana jatamansi* plant.

## 5.5 SUMMARY

According to the WHO fact sheet, depression is a common mental illness worldwide. There is a plethora of scientific studies depicting the promising capabilities of several medicinal plants and their active phytoconstituents. This chapter discusses some of the Himalayan nuts, beans, and grains, along with their nutraceutical phytoconstituents recognized for their therapeutic potential as antidepressants. Many of these studies have evaluated the efficacy of solvent extracts and purified components from leaves, roots, or fruits, and determined their role as antidepressants based on preclinical testing using animal models and experimental assays.

The scientific evidence also suggests the possible application of these herbal compounds in restoring other neurodegenerative disorders such as Alzheimer's, Parkinson's disease, and Dysthymia. However, the underlying mechanisms of the antidepressant action of several of these nuts, grains, and beans remain inconclusive and need further experimentation and exploration.

## ACKNOWLEDGMENT

The authors would like to express their gratitude to Graphic Era Deemed to be University, Dehradun for their help and support during this research.

## KEYWORDS

- **antidepressant**
- **antidepressant activity**
- **antidepressant nuts**
- **bioactive compounds**
- **herbal plants**
- **Himalaya**
- **Himalayan foods**
- **state of mind**

## REFERENCES


1. Abbas, A., Ikram, R., Hasan, F., Adil, A., Nisar, U., & Ain, Q., (2020). Antidepressant and antiamnesic potential of *Foeniculum vulgare*. *Journal of Advances in Medicine and Medical Research, 32*, 131–138.
2. Abozed, S. S., El-Kalyoubi, M., Abdelrashid, A., & Salama, M. F., (2014). Total phenolic contents and antioxidant activities of various solvent extracts from whole wheat and bran. *Annals of Agricultural Sciences, 59*(1), 63–67.
3. Alok, S., Jain, S. K., Verma, A., Kumar, M., Mahor, A., & Sabharwal, M., (2013). Plant profile, phytochemistry and pharmacology of *Asparagus racemosus* (Shatavari): A review. *Asian Pacific Journal of Tropical Disease, 3*(3), 242–251.
4. Anand, A., Khurana, R., Wahal, N., Mahajan, S., Mehta, M., Satija, S., & Khurana, N., (2019). Vanillin: A comprehensive review of pharmacological activities. *Plant Archives, 19*(2), 1000–1004.
5. Arab, L., Guo, R., & Elashoff, D., (2019). Lower depression scores among walnut consumers in NHANES. *Nutrients, 11*(2), 275.
6. Arya, A. K., Arora, M., & Singh, M. F., (2020). A review on pharmacological activity of *Juglans regia*. *International Journal of Pharmacognosy and Chemistry, 7*, 1–11.
7. Devi, P. B., Vijayabharathi, R., Sathyabama, S., Malleshi, N. G., & Priyadarisini, V. B., (2014). Health benefits of finger millet (*Eleusine coracana* L.) polyphenols and dietary fiber: A review. *J. Food Sci. Technol., 51*(6), 1021–1040.
8. Baraniak, J., & Kania-Dobrowolska, M., (2022). The dual nature of amaranth—Functional food and potential medicine. *Foods, 11*(4), 618.
9. Barua, C. C., Haloi, P., Saikia, B., Sulakhiya, K., Pathak, D. C., Tamuli, S., & Ren, X., (2018). *Zanthoxylum alatum* abrogates lipopolysaccharide-induced depression-like behaviors in mice by modulating neuroinflammation and monoamine neurotransmitters in the hippocampus. *Pharmaceutical Biology, 56*(1), 245–252.
10. Barua, U., Gogoi, B., & Deka, K., (2021). *Myrica esculenta* Buch. & Ham. (*Myricaceae*). In: Waman, A. A., & Bohra, P., (eds.), *Perennial Underutilized Horticultural Species of India* (pp. 226–235). Jaya Publishing House, Delhi, India.
11. Bhangale, J. O., & Acharya, S. R., (2016). Anti-Parkinson activity of petroleum ether extract of *Ficus religiosa* (L.) leaves. *Advances in Pharmacological Sciences, 2016.* https://doi.org/10.1155/2016/9436106.
12. Biswas, T. K., & Mukherjee, B., (2003). Plant medicines of Indian origin for wound healing activity: A review. *The International Journal of Lower Extremity Wounds, 2*(1), 25–39.
13. Bloch, R. G., Dooneief, A. S., Buchberg, A. S., & Spellman, S., (1954). The clinical effect of isoniazid and iproniazid in the treatment of pulmonary tuberculosis. *Annals of Internal Medicine, 40*(5), 881–900.
14. Boas, G. R. V., De Lacerda, R. B., Paes, M. M., Gubert, P., Da Cruz Almeida, W. L., Rescia, V. C., et al., (2019). Molecular aspects of depression: A review from neurobiology to treatment. *European Journal of Pharmacology, 851*, 99–121.
15. Braniste, V., Al-Asmakh, M., Kowal, C., Anuar, F., Abbaspour, A., Toth, M., & Pettersson, S., (2014). The gut microbiota influences blood-brain barrier permeability in mice. *Science Translational Medicine, 6*(263), 263ra158–263ra158.

16. Butterweck, V., & Schmidt, M., (2007). St. John's wort: Role of active compounds for its mechanism of action and efficacy. *Wiener Medizinische Wochenschrift*, *157*(13), 356–361.
17. Câmara, C. R., Urrea, C. A., & Schlegel, V., (2013). Pinto beans (*Phaseolus vulgaris* L.) as a functional food: Implications on human health. *Agriculture*, *3*(1), 90–111.
18. Chaturvedi, K., Vishwakarma, D. K., & Singh, N., (2021). COVID-19 and its impact on education, social life, and mental health of students: A survey. *Children and Youth Services Review*, *121*, 105866.
19. Chauhan, A., & Chauhan, V., (2020). Beneficial effects of walnuts on cognition and brain health. *Nutrients*, *12*(2), 550.
20. Cropley, M., Cave, Z., Ellis, J., & Middleton, R. W., (2002). Effect of kava and valerian on human physiological and psychological responses to mental stress assessed under laboratory conditions. *Phytotherapy Research*, *16*(1), 23–27.
21. Dai, W., Feng, K., Sun, X., Xu, L., Wu, S., Rahmand, K., & Han, T., (2022). Natural products for the treatment of stress-induced depression: Pharmacology, mechanism, and traditional use. *Journal of Ethnopharmacology*, *285*, 114692.
22. Delaviz, H., Mohammadi, J., Ghalamfarsa, G., Mohammadi, B., & Farhadi, N., (2017). A review study on phytochemistry and pharmacology applications of *Juglans regia* plant. *Pharmacognosy Reviews*, *11*(22), 145.
23. Devi, M., & Sharma, R., (2014). Antidepressant activity of aqueous extract of *Phaseolus vulgaris* (black bean) in rodent models of depression. *International Journal of Nutrition, Pharmacology, Neurological Diseases*, *4*(2), 118.
24. Devi, P. B., Vijayabharathi, R., Sathyabama, S., Malleshi, N. G., & Priyadarisini, V. B., (2014). Health benefits of finger millet (*Eleusine coracana* L.) polyphenols and dietary fiber: A review. *Journal of Food Science and Technology*, *51*(6), 1021–1040.
25. Dhama, K., Sharun, K., Gugjoo, M. B., Tiwari, R., Alagawany, M., Iqbal, Y. M., & Farag, M. R., (2021). A comprehensive review on chemical profile and pharmacological activities of *Ocimum basilicum*. *Food Reviews International*, 1–29.
26. Dhamija, I., Parle, M., & Kumar, S., (2017). Antidepressant and anxiolytic effects of *Garcinia indica* fruit rind via monoaminergic pathway. *3 Biotech*, *7*(2), 1–12.
27. Dida, G., & Etisa, D., (2019). Effect of lime and compost application on the growth and yield of common bean (*Phaseolus vulgaris* L.): A review. *Ad Oceanogr and Marine Biol.*, *1*(3), 2019.
28. Dinamarca, M. C., Arrazola, M., Toledo, E., Cerpa, W. F., Hancke, J., & Inestrosa, N. C., (2008). Release of acetylcholinesterase (AChE) from β-amyloid plaques assemblies improves the spatial memory impairments in APP-transgenic mice. *Chemico-Biological Interactions*, *175*(1–3), 142–149.
29. Domínguez, F., González-Trujano, E., Gallardo, J. M., & Orozco-Suárez, S., (2016). Antidepressant medicinal plants and compounds used in traditional medicines in North America. In: *Herbal Medicine in Depression* (pp. 381–431). Springer, Cham. NY.
30. El Tanbouly, N., El Sayed, A. M., Ali, Z. Y., Wahab, S. A., El Gayed, S. H., Ezzat, S. M., & Abdel-Sattar, E., (2017). Antidepressant-like effect of selected Egyptian cultivars of flaxseed oil on a rodent model of postpartum depression. *Evidence-based Complementary and Alternative Medicine*, *2017*, 6405789.

31. Erhunmwunsee, F., Pan, C., Yang, K., Li, Y., Liu, M., & Tian, J., (2022). Recent development in biological activities and safety concerns of perillaldehyde from perilla plants: A review. *Critical Reviews in Food Science and Nutrition, 62*(23), 6328–6340.
32. Fajemiroye, J. O., Da Silva, D. M., De Oliveira, D. R., & Costa, E. A., (2016). Treatment of anxiety and depression: Medicinal plants in retrospect. *Fundamental & Clinical Pharmacology, 30*(3), 198–215.
33. Filomeni, G., Graziani, I., De Zio, D., Dini, L., Centonze, D., Rotilio, G., & Ciriolo, M. R., (2012). Neuroprotection of kaempferol by autophagy in models of rotenone-mediated acute toxicity: Possible implications for Parkinson's disease. *Neurobiology of Aging, 33*(4), 767–785.
34. Gaur, P. K., Rastogi, S., & Lata, K., (2022). Correlation between phytocompounds and pharmacological activities of *Boerhavia diffusa* Linn with traditional-ethnopharmacological insights. *Phytomedicine Plus*, 100260.
35. Gohil, K. J., Patel, J. A., & Gajjar, A. K., (2010). Pharmacological review on *Centella asiatica*: A potential herbal cure-all. *Indian Journal of Pharmaceutical Sciences, 72*(5), 546.
36. Goyal, A., Sharma, V., Upadhyay, N., Gill, S., & Sihag, M., (2014). Flax and flaxseed oil: An ancient medicine & modern functional food. *Journal of Food Science and Technology, 51*(9), 1633–1653.
37. Gull, A., Ahmad, N. G., Prasad, K., & Kumar, P., (2016). Technological, processing, and nutritional approach of finger millet (*Eleusine coracana*)-a mini review. *Journal of Food Processing and Technology, 7*(593), 2.
38. Hasin, D. S., Sarvet, A. L., Meyers, J. L., Saha, T. D., Ruan, W. J., Stohl, M., & Grant, B. F., (2018). Epidemiology of adult DSM-5 major depressive disorder and its specifiers in the United States. *JAMA Psychiatry, 75*(4), 336–346.
39. Hayat, I., Ahmad, A., Masud, T., Ahmed, A., & Bashir, S., (2014). Nutritional and health perspectives of beans (*Phaseolus vulgaris* L.): An overview. *Critical Reviews in Food Science and Nutrition, 54*(5), 580–592.
40. Hritcu, L., & Cioanca, O., (2016). Prevalence of use of herbal medicines and complementary and alternative medicine in Europe. In: *Herbal Medicine in Depression* (pp. 135–181). Springer, Cham. NY.
41. Huang, L., Zhao, H., Huang, B., Zheng, C., Peng, W., & Qin, L., (2011). *Acanthopanax senticosus*: Review of botany, chemistry, and pharmacology. *Die Pharmazie-An International Journal of Pharmaceutical Sciences, 66*(2), 83–97.
42. Inestrosa, N. C., Tapia-Rojas, C., Griffith, T. N., Carvajal, F. J., Benito, M. J., Rivera-Dictter, A., & Varela-Nallar, L., (2011). Tetrahydrohyperforin prevents cognitive deficit, Aβ deposition, tau phosphorylation and synaptotoxicity in the APPswe/PSEN1ΔE9 model of Alzheimer's disease: A possible effect on APP processing. *Translational Psychiatry, 1*(7), e20–e20.
43. Islam, M. S., Akhtar, M. M., Rahman, M. M., Rahman, M. A., Sarker, K. K., & Alam, M. F., (2009). Antitumor and phytotoxic activities of leaf methanol extract of *Oldenlandia diffusa* (Willd.) roxb. *Global Journal of Pharmacology, 3*(2), 99–106.
44. Iyer, K., & Khan, J. A., (2012). Depression-a review. *Research Journal of Recent Sciences, 1*(4), 79–87.

45. Joshi, P. V., Shirkhedkar, A. A., Prakash, K., & Maheshwari, V. L., (2011). Antidiarrheal activity, chemical and toxicity profile of *Berberis aristata*. *Pharmaceutical Biology, 49*(1), 94–100.
46. Kabra, A., Baghel, U. S., Hano, C., Martins, N., Khalid, M., & Sharma, R., (2020). Neuroprotective potential of *Myrica esulenta* in haloperidol induced Parkinson's disease. *Journal of Ayurveda and Integrative Medicine, 11*(4), 448–454.
47. Kabra, A., Martins, N., Sharma, R., Kabra, R., & Baghel, U. S., (2019). *Myrica esculenta* Buch.-Ham. ex D. Don: A natural source for health promotion and disease prevention. *Plants, 8*(6), 149.
48. Koley, T. K., Kaur, C., Nagal, S., Walia, S., & Jaggi, S., (2016). Antioxidant activity and phenolic content in genotypes of Indian jujube (*Zizyphus mauritiana* Lamk.). *Arabian Journal of Chemistry, 9*, S1044–S1052.
49. Kris-Etherton, P. M., (2014). Walnuts decrease risk of cardiovascular disease: A summary of efficacy and biologic mechanisms. *The Journal of Nutrition, 144*(4), 547S–554S.
50. Kulkarni, S. K., & Dhir, A., (2008). On the mechanism of antidepressant-like action of berberine chloride. *European Journal of Pharmacology, 589*(1–3), 163–172.
51. Kumar, B. A., Lakshman, K., Velmurugan, C., Sridhar, S. M., & Gopisetty, S., (2014). Antidepressant activity of methanolic extract of *Amaranthus spinosus*. *Basic and Clinical Neuroscience, 5*(1), 11.
52. Kumar, D., & Sharma, P. K., (2021). Traditional use, phytochemicals, and pharmacological activity of *Salvadora persica*: A review. *Current Nutrition & Food Science, 17*(3), 302–309.
53. Kumari, P., Joshi, G. C., & Tewari, L. M., (2011). Diversity and status of ethno-medicinal plants of Almora district in Uttarakhand, India. *International Journal of Biodiversity and Conservation, 3*(7), 298–326.
54. Li, R., Wang, Z. M., Wang, Y., Dong, X., Zhang, L. H., Wang, T., & Xu, Y. T., (2021). Antidepressant activities and regulative effects on serotonin transporter of *Nardostachys jatamansi* DC. *Journal of Ethnopharmacology, 268*, 113601.
55. Lim, S. H., Lee, H. S., Lee, C. H., & Choi, C. I., (2021). Pharmacological activity of *Garcinia indica* (Kokum): An updated review. *Pharmaceuticals, 14*(12), 1338.
56. Lopresti, A. L., (2017). Salvia (Sage): A review of its potential cognitive-enhancing and protective effects. *Drugs in R&D, 17*(1), 53–64.
57. Los, F. G. B., Zielinski, A. A. F., Wojeicchowski, J. P., Nogueira, A., & Demiate, I. M., (2018). Beans (*Phaseolus vulgaris* L.): Whole seeds with complex chemical composition. *Current Opinion in Food Science, 19*, 63–71.
58. Mahmudah, R. A., Hasanuddin, S., Saleh, A., Yuliastri, W. O., & Isrul, M., (2019). Antidepressant activity and identification of chemical compounds extract mustard leaves (*Brassica juncea* L.). *Research Journal of Pharmacy and Technology, 12*(7), 3223–3227.
59. Mamedov, N., & Mamadalieva, N., (2016). Medicinal plants from countries of former USSR used for treatment of depression. In: *Herbal Medicine in Depression: Traditional Medicine to Innovative Drug Delivery* (pp. 183–258). Springer, Cham. NY.
60. Manyam, B. V., Dhanasekaran, M., & Hare, T. A., (2004). Neuroprotective effects of the anti-Parkinson drug *Mucuna pruriens*. *Phytotherapy Research: An International*

*Journal Devoted to Pharmacological and Toxicological Evaluation of Natural Product Derivatives*, *18*(9), 706–712.

61. Marder, M., Viola, H., Wasowski, C., Fernández, S., Medina, J. H., & Paladini, A. C., (2003). 6-methylapigenin and hesperidin: New *Valeriana* flavonoids with activity on the CNS. *Pharmacology Biochemistry and Behavior*, *75*(3), 537–545.
62. Moloudizargari, M., Mikaili, P., Aghajanshakeri, S., Asghari, M. H., & Shayegh, J., (2013). Pharmacological and therapeutic effects of *Peganum harmala* and its main alkaloids. *Pharmacognosy Reviews*, *7*(14), 199.
63. Motti, R., & De Falco, B., (2021). Traditional herbal remedies used for managing anxiety and insomnia in Italy: An ethnopharmacological overview. *Horticulturae*, *7*(12), 523.
64. Nagori, K., Singh, M. K., Alexander, A., Kumar, T., Dewangan, D., Badwaik, H., & Tripathi, D. K., (2011). Piper betle L.: A review on its ethnobotany, phytochemistry, pharmacological profile and profiling by new hyphenated technique DART-MS (direct analysis in real time mass spectrometry). *Journal of Pharmacy Research*, *4*(9), 2991–2997.
65. Nouri, F., Sarrafzadegan, N., Mohammadifard, N., Sadeghi, M., & Mansourian, M., (2016). Intake of legumes and the risk of cardiovascular disease: Frailty modeling of a prospective cohort study in the Iranian middle-aged and older population. *European Journal of Clinical Nutrition*, *70*(2), 217–221.
66. Oliveira, A. I., Pinho, C., Sarmento, B., & Dias, A. C., (2016). Neuroprotective activity of *Hypericum perforatum* and its major components. *Frontiers in Plant Science*, *7*, 1004.
67. Oosterwijk, S., Lindquist, K. A., Anderson, E., Dautoff, R., Moriguchi, Y., & Barrett, L. F., (2012). States of mind: Emotions, body feelings, and thoughts share distributed neural networks. *Neuro Image*, *62*(3), 2110–2128.
68. Pal, R. S., Kumar, R. A., Kant, L., & Bhatt, J. C., (2014). Kilmora, a wild edible potential nutraceutical fruit in Indian Himalayan Region. *Popular Kheti*, *2*(3), 199–203.
69. Palazidou, E., (2012). The neurobiology of depression. *British Medical Bulletin*, *101*(1), 127–145.
70. Pan, A., Sun, Q., Manson, J. E., Willett, W. C., & Hu, F. B., (2013). Walnut consumption is associated with lower risk of type 2 diabetes in women. *The Journal of Nutrition*, *143*(4), 512–518.
71. Pandareesh, M. D., Chauhan, V., & Chauhan, A., (2018). Walnut supplementation in the diet reduces oxidative damage and improves antioxidant status in transgenic mouse model of Alzheimer's disease. *Journal of Alzheimer's Disease*, *64*(4), 1295–1305.
72. Pandey, D. K., Rajkumar, R., Mahesh, R., & Radha, R., (2008). Depressant-like effects of parthenolide in a rodent behavioral antidepressant test battery. *Journal of Pharmacy and Pharmacology*, *60*(12), 1643–1650.
73. Pati, D., Pandey, D. K., Mahesh, R., Kurdekar, V., & Jhadav, H. R., (2010). Anti-depressant-like activity of *Mucuna pruriens*; A traditional Indian herb in rodent models of depression. *Phamacology Online*, *1*, 537–551.

74. Petimani, M. S., Adake, P., & Aditya, A., (2019). Evaluation of antidepressant activity of *Dolichos* biflorus methanolic seed extract in animal models of depression. *International Journal of Biochemistry and Biophysics*, *7*(1), 1–5.
75. Poorbaferani, F., Rouhani, M. H., Heidari, Z., Poorbaferani, M., & Safavi, S. M., (2020). Flaxseed oil supplementation on severity of depression and brain-derived neurotrophic factor: A randomized, double blind placebo controlled clinical trial. *International Journal of Food Properties*, *23*(1), 1518–1526.
76. Poornachandra, M. N., Khanam, S., Shivananda, B. G., Shivananda, T. N., & Dris, R., (2005). *Mucuna pruriens* (L.) DC-A novel drug for learning and memory retrieval. *International Journal of Food, Agriculture and Environment,* 3, 13–15.
77. Popli, S. P., (1968). Medicinal plants chemical constituents of *Berberis aristata. Indian Journal of Chemistry & Biodiversity*, *6*, 123.
78. Popoveniuc, B., (2014). The philosophy of depression. *International Journal of Communication Research, 4*(1), 22–31.
79. Pragya, S., & Raghuvanshi, S. R., (2012). *Finger millet* for food and nutritional security. *African Journal of Food Science*, *6*(4), 77–84.
80. Prakash, O., Usmani, S., Singh, R., Singh, N., Gupta, A., & Ved, A., (2021). A panoramic view on phytochemical, nutritional, and therapeutic attributes of *Ziziphus mauritiana* Lam.: A comprehensive review. *Phytotherapy Research*, *35*(1), 63–77.
81. Pribis, P., Bailey, R. N., Russell, A. A., Kilsby, M. A., Hernandez, M., Craig, W. J., & Sabate, J., (2012). Effects of walnut consumption on cognitive performance in young adults. *British Journal of Nutrition*, *107*(9), 1393–1401.
82. Ramnath, S., & Suri, G., (2021). Managing depression in India: Opportunities for a targeted smartphone app. *International Journal of Social Psychiatry*, *67*(8), 1035–1045.
83. Rana, R. K., & Patel, R. K., (2014). Antioxidant activity of bark of *Myrica Nagi. International Journal of Pharmaceutical Sciences Review and Research*, *28*(1), 99–101.
84. Ravi, S., Jyothi, P., Shanmugam, B., Subbaiah, G. V., Prasad, S. H., & Reddy, K. S., (2021). A comprehensive review on traditional knowledge, phytochemistry and pharmacological properties of *Acalypha indica* L. *Pharmacognosy Reviews*, *15*(30).
85. Rawat, S., Kumar, N., & Kothiyal, P., (2013). Evaluate the antidiabetic activity of *Myrica esculenta* leaves in streptozotocin induced diabetes in rat. *International Journal of Universal Pharmacy and Bio Sciences, 2*, 510–525.
86. Razavi, B. M., Hosseini, A., & Hosseinzadeh, H., (2021). Antidepressant activity of *Crocus sativus* L. and its main constituents: A review. *The Neuroscience of Depression*, 493–502.
87. Rook, G. A., Lowry, C. A., & Raison, C. L., (2013). Microbial 'old friends,' immunoregulation and stress resilience. *Evolution, Medicine, and Public Health, 2013*(1), 46–64.
88. Russo, E., Scicchitano, F., Whalley, B. J., Mazzitello, C., Ciriaco, M., Esposito, S., & De Sarro, G., (2014). *Hypericum perforatum*: Pharmacokinetic, mechanism of action, tolerability, and clinical drug–drug interactions. *Phytotherapy Research*, *28*(5), 643–655.

89. Sachan, N., Shrivastav, A., & Pal, D., (2019). Evaluation of antidepressant activity of ethanolic extract of *Abies webbiana* and *Berberis aristata* in laboratory animals. *Journal of Drug Delivery and Therapeutics*, *9*(1), 244–247.
90. Maharajan, T., Antony, C. S., Ajeesh, K. T. P., & Ignacimuthu, S., (2021). Finger millet [*Eleusine coracana* (L.) Gaertn]: An orphan crop with a potential to alleviate the calcium deficiency in the semi-arid tropics of Asia and Africa. *Front. Sustain. Food Syst.*, *5*, 684447.
91. Samundeeswari, C., Rajadurai, M., Periasami, R., & Kanchana, G., (2011). Hepatoprotective effect of herbitars, A polyherbal against $CCl_4$ induced hepatotoxicity in rats. *Journal of Pharmacy Research*, *4*, 676–679.
92. Sarikurkcu, C., Ozer, M. S., & Tlili, N., (2019). LC–ESI–MS/MS characterization of phytochemical and enzyme inhibitory effects of different solvent extract of *Symphytum anatolicum*. *Industrial Crops and Products*, *140*, 111666.
93. Schmidt, M., & Mellon, D., (2010). Neuronal processing of chemical information in crustaceans. In: *Chemical Communication in Crustaceans* (pp. 123–147). Springer, New York, NY.
94. Shahidi, F., & Naczk, M., (2003). *Phenolics in Food and Nutraceuticals*. CRC press.
95. Shao, J., Wei, Y., & Wei, X., (2022). A comprehensive review on bioavailability, safety, and antidepressant potential of natural bioactive components from tea. *Food Research International*, 111540.
96. Sharma, R., (2022). A review on therapeutic intervention of yoga and ayurveda in post-partum depression. *Open Journal of Obstetrics and Gynecology*, *12*(3), 201–208.
97. Sherwin, E., Sandhu, K. V., Dinan, T. G., & Cryan, J. F., (2016). May the force be with you: The light and dark sides of the microbiota–gut–brain axis in neuropsychiatry. *CNS Drugs*, *30*(11), 1019–1041.
98. Shobana, S., Krishnaswamy, K., Sudha, V., Malleshi, N. G., Anjana, R. M., Palaniappan, L., & Mohan, V., (2013). Finger millet (Ragi, *Eleusine coracana* L.): A review of its nutritional properties, processing, and plausible health benefits. *Advances in Food and Nutrition Research*, *69*, 1–39.
99. Shoeb, A., Chowta, M., Pallempati, G., Rai, A., & Singh, A., (2013). Evaluation of antidepressant activity of vanillin in mice. *Indian Journal of Pharmacology*, *45*(2), 141.
100. Shukla, A., (2021). Ethnic food culture of Chhattisgarh state of India. *Journal of Ethnic Foods*, *8*(1), 1–16.
101. Siddiqui, M. J., Saleh, M. S., Basharuddin, S. N. B., Zamri, S. H. B., Bin Mohd, N. M. H., Bin Che, I. M. Z., & Khatib, A., (2018). Saffron (*Crocus sativus* L.): As an antidepressant. *Journal Of Pharmacy & Bioallied Sciences, 10*(4), 173.
102. Singh, T. P., & Singh, O. M., (2011). Phytochemical and pharmacological profile of *Zanthoxylum armatum* DC.: An overview. *Indian Journal of Natural Product and Research*, *2*(3), 275–285.
103. Sood, P., & Shri, R., (2018). A review on ethnomedicinal, phytochemical and pharmacological aspects of *Myrica esculenta*. *Indian Journal of Pharmaceutical Sciences*, *80*(1), 2–13.
104. Suneja, Y., Kaur, S., Gupta, A. K., & Kaur, N., (2011). Levels of nutritional constituents and antinutritional factors in black gram (*Vigna mungo* L. Hepper). *Food Research International*, *44*(2), 621–628.

105. Taiwo, A. E., Leite, F. B., Lucena, G. M., Barros, M., Silveira, D., Silva, M. V., & Ferreira, V. M., (2012). Anxiolytic and antidepressant-like effects of *Melissa officinalis* (lemon balm) extract in rats: Influence of administration and gender. *Indian Journal of Pharmacology*, *44*(2), 189.
106. Teodoro, A. J., (2019). Bioactive compounds of food: Their role in the prevention and treatment of diseases. *Oxidative Medicine and Cellular Longevity*, *2019*. https://doi.org/10.1155/2019/3765986.
107. Thakur, A. K., Chatterjee, S. S., & Kumar, V., (2014). Antidepressant-like effects of *Brassica juncea* L. leaves in diabetic rodents. *Indian Journal of Experimental Biology*, *52*, 613–622.
108. Tikiyani, V., & Babu, K., (2019). Claudins in the brain: Unconventional functions in neurons. *Traffic*, *20*(11), 807–814.
109. Tindall, A. M., Petersen, K. S., Skulas-Ray, A. C., Richter, C. K., Proctor, D. N., & Kris-Etherton, P. M., (2019). Replacing saturated fat with walnuts or vegetable oils improves central blood pressure and serum lipids in adults at risk for cardiovascular disease: A randomized controlled-feeding trial. *Journal of the American Heart Association*, *8*(9), e011512.
110. Valls-Pedret, C., Sala-Vila, A., Serra-Mir, M., Corella, D., De La Torre, R., Martínez-González, M. Á., & Ros, E., (2015). Mediterranean diet and age-related cognitive decline: A randomized clinical trial. *JAMA Internal Medicine*, *175*(7), 1094–1103.
111. Warrier, P. K., (1993). *Indian Medicinal Plants: A Compendium of 500 Species* (Vol. 5). Orient Blackswan.
112. WHO Director-General's Opening Remarks at Mental Health Side Event, Commonwealth Heads of Government Meeting, Rwanda, (2022). *Rethinking Mental Health: A Commonwealth Call to Support, Care and Transform*.
113. Wink, M., Botschen, F., Gosmann, C., Schäfer, H., & Waterman, P. G., (2010). Chemotaxonomy seen from a phylogenetic perspective and evolution of secondary metabolism. *Annual Plant Reviews: Biochemistry of Plant Secondary Metabolism*, *40*, 364–433.
114. Wurglics, M., & Schubert-Zsilavecz, M., (2006). Hypericum perforatum: A 'modern' herbal antidepressant. *Clinical Pharmacokinetics*, *45*(5), 449–468.
115. Yahia, Y., Benabderrahim, M. A., Tlili, N., Bagues, M., & Nagaz, K., (2020). Bioactive compounds, antioxidant, and antimicrobial activities of extracts from different plant parts of two *Ziziphus* Mill. species. *PloS One*, *15*(5), e0232599.
116. Zhang, L., Wang, L., Huang, L., Zhao, Y., Ding, H., Li, B., Wen, L., Liu, Y., et al., (2021). Antidepressant effects of total iridoids of *Valeriana jatamansi* via the intestinal flora-blood–brain barrier pathway. *Pharmaceutical Biology*, *59*(1), 910–919.
117. Zhao, J. L., Jiang, W. T., Wang, X., Cai, Z. D., Liu, Z. H., & Liu, G. R., (2020). Exercise, brain plasticity, and depression. *CNS Neuroscience & Therapeutics*, *26*(9), 885–895.

# CHAPTER 6

# Antidepression Agents in Himalayan Mushrooms

TAJ-UD-DIN,[1] MUHAMMAD RIZWAN,[1]
MOHAMMAD NAZIR UDDIN,[1] MUHAMMAD FAISAL SIDDIQUI,[2]
ABDUR RAUF,[3] ABDULLAH S. M. ALJOHANI,[4] and
WALED AL ABDULMONEM[5]

[1]*Center for Biotechnology and Microbiology, University of Swat, Swat, Tehsil, Charbagh, Khyber Pakhtunkhwa, Pakistan*

[2]*Department of Microbiology, Hazara University, Mansehra, Pakistan*

[3]*Department of Chemistry, University of Swabi, Swabi, Anbar, Khyber Pakhtunkhwa, Pakistan*

[4]*Department of Veterinary Medicine, College of Agriculture and Veterinary Medicine, Qassim University, Buraydah, Saudi Arabia*

[5]*Department of Pathology, College of Medicine, Qassim University, Buraydah, Saudi Arabia*

## ABSTRACT

The Himalayan region is rich in flora and fauna. Various species of mushrooms are found in the Himalayan regions. The history of the use of wild mushrooms as food and in medicine is well-documented for many countries. Natural antibiotics are abundant in mushrooms, and the secondary metabolites have been shown to have antibacterial and antiviral activity. Exudates from mushroom mycelia are active against protozoans. Mushrooms that contain psilocybin, popularly known as psychedelic mushrooms, have psychedelic qualities. Psychedelic mushrooms are used

Himalayan Medicinal Plants for the Treatment of Depression: A Source of Rich Antidepressant Agents.
Abdur Rauf, Prabhakar Semwal, & Hafiz Ansar Rasul Suleria (Eds.)

for treating depression. Psilocybin in mushrooms temporarily disrupts brain activity, producing a sort of "reset," which enhances neuron transmission and communication, returns neurotransmitter levels to normal, and improves blood supply to brain regions that are underactive. As with all interventions in health, there are benefits and drawbacks to treating depression with psilocybin. To evaluate the true potential and availability of mushroom species and to optimize manufacturing processes to increase the quality and quantity of antidepressant compounds present in mushrooms, more thorough study is required. To develop a rapid propagation protocol to commercially yield high-demanding mushroom species, to breed improved mushroom variants with high psychedelic yields, and to cultivate mushrooms for the manufacturing of antidepressants and other industrial applications, bioengineering and other modern techniques are necessary.

## 6.1 INTRODUCTION

Mushrooms are the fruiting structures of certain Ascomycetes and Basidiomycetes, characterized by recognizable carpophores [42]. They are cultivated worldwide for food and medicinal purposes, as well as found in the wild [10]. Research on their diversity in Pakistan has identified 23 species of ectomycorrhizal fungi from 11 genera [39]. *Inocybe* and *Hymenoscyphus* have been identified as dominant mushroom species [39]. Fungi species used in food and cooking have been discovered through morphological and genetic methods [50]. Previous diversity analyses of fungi have relied on conventional techniques [2]. Targeted surveys have proven more effective than random surveys in identifying mushroom species [38]. Plot-based diversity approaches have been employed to gather data on the baseline fungal community. The distribution and diversity of fungi have also been studied using the quadrat approach [113]. The line transect approach can be used to compare different fungal communities, conserve species, and gain insights into the factors influencing the structure and connectivity of fungal communities. Information on the timing of fungal development and maturation is also provided [1].

Mushrooms are used in many traditional treatments and have various health-promoting properties [49, 67, 103]. A recent field of study called ethnomycology is devoted to how fungi interact with regional societies. It covers traditional, recreational, and cultural applications of mushrooms [93]. Mushrooms are a naturally renewable and underutilized resource that helps improve the standard of living in rural areas. They are becoming increasingly important among scientific and research groups worldwide due to their wide-ranging ecological, pharmacological, nutritional, and health-promoting features [80]. Non-timber forest products (NTFPs) like wild mushrooms are gathered for food and income [29]. Different cultures have varied practices for gathering and using wild edible mushrooms (WEM) [95]. These mushrooms are collected and sold for food and other purposes. Folk taxonomic research on fungi is crucial since several mushroom species are on the verge of extinction [54]. It is beneficial to pass along traditional mycological knowledge from one generation to the next in order to protect the uses and applications of edible mushrooms [52].

In the world, and especially in rural areas of Pakistan, morels are a crucial source of both food and income [92]. They are used in traditional treatments for common illnesses. To enhance mushroom cultivation, utilization, and conservation among ethnic mountain communities, it is important to share traditional knowledge about mushrooms [64].

The entire Himalayan region is blessed with diverse climatic and geographic conditions, which support a wide range of mushroom species [18]. Despite the presence of numerous ethnic groups in the region, there is a lack of well-documented ethno-mycological data, and no comprehensive studies have been conducted to explore these resources for human welfare [90]. The diversity, specific habitats, ethno-mycological applications, production, harvesting, and export of mushrooms are all poorly documented. The objective of this chapter is to document the mushroom varieties found in the Himalayan region and their use as an anti-depressant agent.

## 6.2 ETYMOLOGY

Both the definition and usage of the words "mushroom" and "toadstool," which have been around for millennia, have never been agreed upon. In the 15$^{th}$ and 16$^{th}$ centuries, people used the words "musshrom," "musshrum," "musscheron," "mousheroms," "mussheron," or "musserouns" [4]. The word "mushroom" and its derivatives come from the word "mousseron," a French term denoting moss (mousse) [65]. There are several types of mushrooms, some of which may be edible, toxic, or unappealing, as the difference between poisonous and edible fungus is not always clear [55, 63]. The term "toadstool" initially arose in English in the 14$^{th}$ century to refer to a "stool" for toads, possibly suggesting a toxic, inedible fungus [43].

## 6.3 IDENTIFICATION

Mushroom identification requires a fundamental understanding of its macroscopic features. The majority of mushrooms belong to *Basidiomycota.*

Basidiospores, which are produced on the gills, are responsible for the release of a fine powdery substance from beneath the caps. Microscopically, basidiospores settle into empty spaces between the gills after being expelled from the basidia. For this reason, when the cap of most mushrooms is removed and placed gill-side down on the ground for an extended period of time, a powdery impression resembling the shape of the gills (spines or pores) will remain. The color of the powdery print, also known as the spore print, helps classify and identify different types of mushrooms. In addition to white spots, there are various brown spots, black dots, a dark purple-brown spot, a pink spot, a yellow spot, and a creamy spot; blue, green, and red spots are rare to find [31, 37] (Figures 6.1 and 6.2).

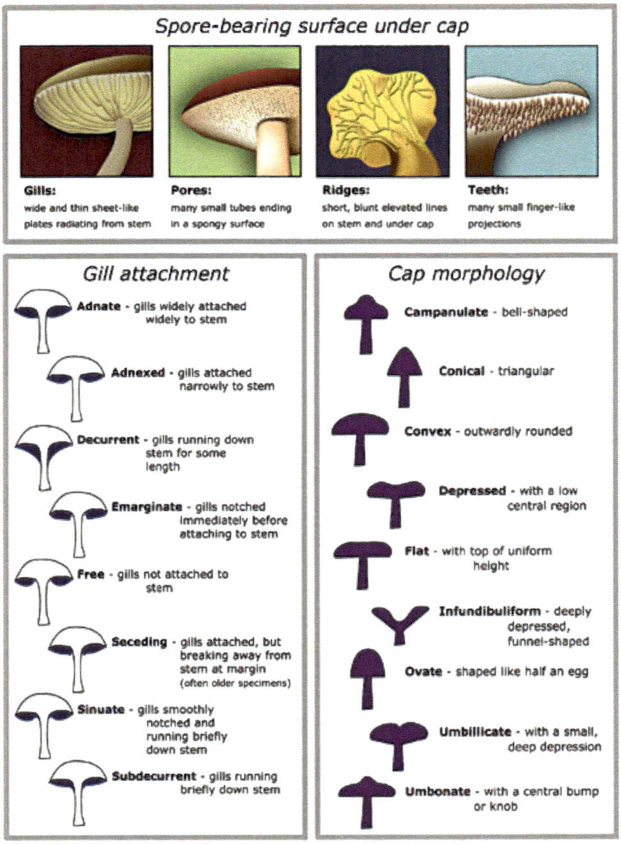


**FIGURE 6.1** Features of mushroom's cap morphology and gills attachment.

*Source*: Reprinted from https://en.wikipedia.org/wiki/File:Mushroom_cap_morphology2.png. https://creativecommons.org/licenses/by-sa/3.0/deed.en

**FIGURE 6.2** A polypore mushroom called *maitake*.

*Source*: Photo by Patrick Harvey. https://creativecommons.org/licenses/by-sa/3.0/deed.en

Although most people still employ traditional methods for identification, which have evolved into a fine art with roots in the Victorian era and medieval ages, along with microscopic study, current mushroom identification is fast moving toward molecular techniques. Both novices and experienced mycologists need to consider several variables when discussing fungi, such as fluids that appear upon disturbance, tastes, shades of color and odors, as well as season, habit, and habitat. Due to toxins and allergies, eating and smelling mushrooms comes with its own set of risks. For several genera, chemical testing is also employed [59].

In general, using a local mushroom genus identification guide may frequently be sufficient. However, species documentation requires more work. The only part of a mushroom that may display particular characteristics needed for species identification is the adult structure, which develops from the button stage. However, specimens that are over-mature tend to degrade and stop producing spores. Many novices have misidentified white spore prints as paper stains from moist water marks or colorful spore prints as paper stains from liquid drips on lamella edges [91].

## 6.4 CLASSIFICATION

*Agaricus campestris* is the type species and genus of the *Agaricales* order, which includes typical mushrooms. There are numerous species of gilled fungi in the genus *Agaricomycetes,* including many that produce fruiting bodies. However, according to the current molecular classifications, not all members of this order produce mushroom fruiting bodies. In addition, there are numerous other gilled fungi in several orders of the class *Agaricomycetes* that produce mushroom fruiting bodies. There is a broad range of mushroom species in the *Cantharellales* order, including *chanterelles, imitation chanterelles,* milkcap mushrooms (*Lactarius, Lactifluus*), and russulas (*Russula*), which includes the hardy, leathery genus *Lentinus* and *Panus*, *Neolentinus* is a member of *Gloeophyllales,* and the tiny genus *pin-mushroom* [17, 91] (Figure 6.3).

**FIGURE 6.3** *Hypomyces lactifluorum* parasitizes *Russula brevipes*, creating a lobster mushroom.

*Source*: Photo by Alan Rockefeller. https://creativecommons.org/licenses/by-sa/4.0/deed.en

There are a number of common fungi that belong to the *Agaricales* category, including mushroom species such as the wild enoki, fly agaricus, fairy-ring mushroom, oyster mushrooms, shiitake, and various *Amanitas,* as well as hallucinogenic mushrooms such as *Psilocybe* species, shaggy manes, and paddy straw mushrooms [107].

*Hypomyces lactifluorum* is a parasitic *ascomycete* which has colored and distorted the fruit-body of *Russula* and *Lactarius* that is parasitized by the mycoparasitic ascomycete *Hypomyces lactifluorum* [100] (Figure 6.4).

**FIGURE 6.4** Lobster mushroom.

*Source*: Photo by Joe Mabel. https://creativecommons.org/licenses/by-sa/3.0/

The term "mushroom" is used ambiguously when referring to non-gilled mushrooms, making it difficult to fully categorize them. Some of these mushrooms have holes on the underside and are commonly known as boletes. Another method of spore dispersal is through teeth, also known as spines. Tooth fungi are characterized by long, thin teeth that hang downward and release spores. They function similarly to gilled fungi, but with the spore release occurring through their spiny teeth. Compared to gilled and pored mushrooms, toothed mushrooms are less common but still encompass some interesting species. One such species is the hedgehog mushroom (*Hydnum repandum*), which is closely related to the edible golden chanterelle. Another species, *Hericium erinaceus*, is a tasty edible mushroom with potential medicinal benefits. It is also known as Lion's Mane, Bearded Tooth, and Bearded Hedgehog. As a result, the term "mushroom" is commonly used to refer to macroscopic fungal fruiting structures rather than having a clear taxonomic definition. There are approximately 14,000 different species of mushrooms known [110].

## 6.5 MORPHOLOGY

The primordium, also known as a nodule or pinhead with a diameter of less than two millimeters, usually grows on or close to the substrate's surface. It develops inside the mycelia, a collection of slender hyphae that constitute the fungus. A "button"-shaped structure of intertwined hyphae that approximately resembles an egg develops from the primordium as it enlarges. The global veil, which is made out of a cottony roll of mycelium, surrounds the growing fruiting bodies in the button. The universal covering tears as the egg grows, and it may stay as a cup. Numerous mushrooms have no universal covering, which prevents them from having a volva or volval patches. The blade-like gills that carry spores are frequently covered by a secondary layer of tissues called the partial covering. An annulus or ring might appear around the center of the stalk as the cap expands, or there may be pieces of the fractional veil hanging from the cap's edge. The ring may resemble a skirt, as in certain Amanita species, a collar, as in many *Lepiota* species, or it may only be traces of a Cortina, a fractional veil comprised of filaments that mimic a spiderweb and is unique to the genus *Cortinarius*. Mushrooms that are partially obscured don't form an annulus [83, 107] (Figures 6.5 and 6.6).

**FIGURE 6.5** *Amanita jacksonii* emerging from veils.

*Source*: Photo by hríb. https://creativecommons.org/licenses/by-sa/3.0/legalcode

**FIGURE 6.6** A milk-cap mushroom *Lactarius indigo.*

*Source*: Photo by Dan Molter. https://creativecommons.org/licenses/by-sa/3.0/

The stalk, which is also known as the stipe or stem, may be lateral or off-center in some species of *Panus* and *Pleurotus,* supporting the cap in the center. Numerous mushrooms do not have a stem, such as those with polypores that resemble brackets with shelves. Although puffballs do not have a stalk, they sometimes do. There is a specialized mycological terminology to describe the structures of other mushrooms, including truffles, jellies, earthstars, and bird's nests, which often lack stems [78].

An important aspect of mushroom morphology is the way the gills attach to the top of the stem. The free gills on mushrooms in the genera *Amanita, Agaricus, Pluteus,* and *Lepiota* do not reach the top of the stalk. Others, like those belonging to the genus *Omphalotus* and *Pleurotus,* have decurrent gills that continue down the stalk. The term "connected gills" refers to the wide range of differences among the variations of free and decurrent gills. The various forms of attached gills are sometimes distinguished by finer characteristics, such as adnate gills, which attach to the stalk squarely, and irregular lamellae, which have a notch that connects them to the stalk [56, 58].

## 6.6 MICROSCOPIC FEATURES

The surface of gills is covered by a hymenium, a layer of tiny spore-bearing cells. The inside surfaces of boletes and polypore tubes, as well as the teeth of spine fungus and the branches of corals in non-gilled mushrooms, are all covered in hymenium. Spores develop within small, elongated cells that resemble sacs in the *Ascomycota*. Typically, an ascus has eight spores [6]. As visible on the inside surfaces of cup fungi or inside the morel pits, the *Discomycetes,* which includes the cup, sponge, brain, and numerous club-like fungi, develop an exposed sheet of asci [24]. The asci develop in tiny, flask-shaped structures known as perithecia, which are made by the *Pyrenomycetes,* a small, dark-colored fungus that survives on a variety of substrata like dung, leaf litter, rotting wood, and soil [22] (Figure 6.7).

A number of four spores often form on the terminal region of sterigmata in *Basidiomycetes,* which emerge from basidia, or club-shaped cells. The fertile region of *Gasteromycetes,* known as a Gleba, can take on different textures, such as powdered puffballs or slimy stinkhorns. Paraphyses are thread-like sterile cells that are scattered amid the asci. The hymenium of the *Basidiomycota* frequently contains comparable structures termed cystidia. There are several kinds of cystidia, and determining a mushroom's

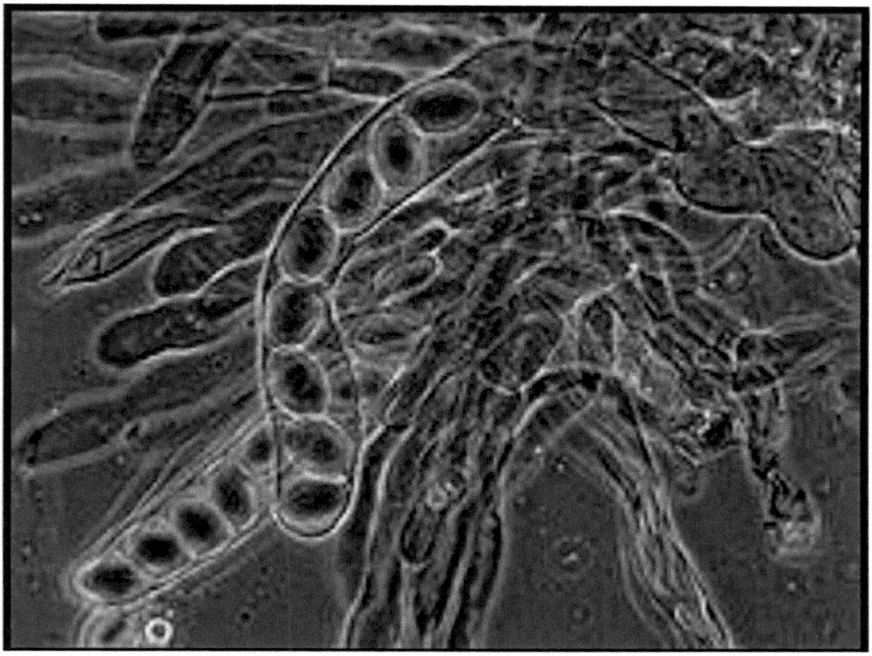

**FIGURE 6.7** Phase contrast microscopy image of *Morchella elata.*

*Source*: Photo by Peter G. Werner. https://creativecommons.org/licenses/by/3.0/deed.en

identity frequently involves examining their number, appearance, and dimensions [57].

The spores are the most crucial microscopic characteristic for identifying mushrooms [30]. Identification is frequently based on characteristics such as color, shape, texture, adherence, ornamentation, responsiveness to chemicals, and absorption [15].

## 6.7 GROWTH

A large variety of mushroom species seem to appear out of nowhere, grow, and spread swiftly. The expressions "to mushroom" or "mushrooming" (growing swiftly in volume or range) and "to pop up like a mushroom" are also derived from this phenomenon (to appear surprisingly and rapidly). The creation of the original fruit bodies of the mushroom really takes many days for all species of mushrooms; however, the process is sped up by the absorption of liquids [28, 106] (Figure 6.8).

**FIGURE 6.8** Emergence of *Agaricus bitorquis* through asphalt.

*Source*: Photo by Tangopaso. Public domain.

The initial fruiting body produced by both common field mushrooms and farmed mushrooms is referred to as the pin stage, named for its small size. When they grow larger, they are called buttons due to their size and shape. Once these stages have emerged, the mushroom-shaped cloud can rapidly expand by inflating prepared cells that have developed primordia over several days. Another ability is the quick absorption of water from its mycelium [20, 75].

Mushrooms, such as *Parasolaplicatilis,* also grow rapidly. The primordia begin to develop at ground level in moist areas, and with significant rainfall or in humid conditions, they quickly reach their maximum size, release spores, and then deflate. They grow to their full size [47]. Some mushrooms grow very slowly, sprouting from the edges of the colony or injecting hyphae to add tissue to their fruiting bodies. Not all species of mushrooms grow quickly. For example, *Pleurotus nebrodensis* is currently highly threatened due to its slow growth and human harvesting [109] (Figure 6.9).

**FIGURE 6.9** *Leucocoprinus birnbaumii* (developmental stages).

*Source*: Photo by Rolf Lawrenz. https://creativecommons.org/licenses/by/4.0/deed.en

The mycelium underlying mushroom fruiting bodies, however, can grow to be very large and have a long lifespan. In the Malheur National Forest in the United States, there is a colony of *Armillaria solidipes* that is believed to be at least 2,400 years old, and possibly even older, covering an area of 2,200 acres [15]. Most of these mushrooms are found growing in decaying tree roots or rotting wood. They appear as black shoelace-like rhizomorphs and white mycelia that connect the colonized surfaces of split timber [12].

## 6.8 NUTRITION

While brown mushrooms are raw, they contain 92% water, 4% carbs, 2% protein, and less than 1% fat. Raw mushrooms have 22 calories per 100 g (3.5 ounces) and are a rich source of B vitamins such as riboflavin, niacin, and pantothenic acid (20% or more of the daily value (DV)), selenium (37% DV), copper (25% DV), and a moderate source of phosphorus, zinc, and potassium (10–19% DV). They are extremely low in salt and vitamin C [85, 105].

### *6.8.1 VITAMIN D*

Post-harvest practices, particularly unintentional sun exposure, have an impact on the vitamin D content of mushrooms. The US Department of Agriculture presented data showing that mushrooms exposed to UV light had high amounts of vitamin D content [13]. Even after harvest, ergosterol in mushrooms is purposely transformed to vitamin D2 when exposed to ultraviolet (UV) light, in order to supply fresh vitamin D mushrooms for the functional food retail market. The effectiveness of artificial UV light technology in creating vitamin D in fresh mushrooms was thoroughly examined by researchers, who found that doing so was just as safe as exposing the mushrooms to sunlight. It has long been considered safe to generate vitamin D in food using UV light [94].

## 6.9 HUMAN USES

Human uses mushrooms for various purposes, some of which are discussed in subsections.

### *6.9.1 EDIBLE MUSHROOMS*

Mushrooms are widely used in a variety of dishes, such as those from Chinese, European, Japanese, and Korean cuisines [35]. The majority of mushrooms traded in stores are mainly harvested with an eye on profit.

The most well-known of these is *Agaricus bisporus,* which is created under sterile, regulated conditions, making it safe for consumption by the majority of humans. The white, cremini, and portobello cultivars of *A. bisporus* are used for commercial production [68]. Along with other cultivated species, many supermarkets also sell Enoki, Pleurotus, Maitake, Shiitake, and *Hericium erinaceus* [98]. Growing awareness of mushroom farming, which is usually considered a possible source of revenue for small farmers, has recently developed as a result of growing wealth in emerging countries [87] (Figures 6.10 and 6.11).

**FIGURE 6.10** *Agaricus bisporus*, widely cultivated mushroom.

*Source*: Photo by chris_73. https://creativecommons.org/licenses/by-sa/3.0/deed.en

A significant producer of edible mushrooms is China. This country produces around half of the world's farmed mushrooms, and each of the world's 1.4 billion inhabitants consumes 2.7 kg (6.0 lb) of mushrooms yearly. Poland is the world's largest exporter of mushrooms, having recorded exports of 194,000 tons in 2014 [60].

It is particularly difficult to distinguish between edible and poisonous species since neither all-dangerous mushrooms nor all edible mushrooms can be identified by a single characteristic. Mushrooming, or simply

**FIGURE 6.11** Various shapes and colors of culinary mushrooms.

*Source*: Photo by Harmonywriter. https://creativecommons.org/licenses/by-sa/3.0/

"hunting for mushrooms," is the practice of gathering them for human consumption. *Mycophagists* are those who engage in this practice. Even edible mushrooms can cause allergic responses in those who are sensitive, ranging from a moderate asthmatic reaction to an extreme instance of anaphylaxis. Agaritine is the most common of the hydrazine found in even the cultivated *A bisporus*. However, heating over a moderate heat destroys the hydrazine [36, 89].

Even though some of them resemble edible species, some mushroom species are deadly, and eating them can be fatal. Gathering wild mushrooms is dangerous and should only be done by someone who is familiar with mushroom identification. Wild mushroom collectors should focus on gathering a few types of recognizable, palatable mushrooms that are difficult to confuse with deadly variations. The general consensus among mushroom hunters is that if a mushroom cannot be positively recognized, it should be assumed to be toxic and should not be consumed [61, 96].

### 6.9.2 *TOXIC MUSHROOMS*

Numerous types of mushrooms create secondary metabolites, some of which are poisonous, hallucinogenic, antibiotic, antiviral, or bioluminescent. Although there aren't many species that can actually be lethal, some of them can still cause incredibly severe symptoms. Toxicology likely plays a role in maintaining the functionality of the basidiocarp, as the mycelium expends a lot of energy and protoplasm to create a structure that effectively spreads its spores. One method of preventing ingestion and early death is the development of ingredients that render the mushroom inedible, forcing the consumer to either vomit the meal or learn to completely avoid eating it. Additionally, research on the potential toxicity of European mushrooms after the Chernobyl incident in 1986 persisted until 2008. Mushrooms are a good illustration of this due to their ability to absorb heavy metals, including radioactive ones [53, 104] (Figure 6.12).

**FIGURE 6.12** Young "death cap" *Amanita phalloides* mushrooms next to a matchbox for scale.

*Source*: Photo by Stanisław Skowron. Public domain.

### 6.9.3 FOLK MEDICINE

Folk medicine utilizes certain mushrooms. Several countries have government-registered extracts for use as adjunct cancer medications, including lentinan, polysaccharide peptide, polysaccharide-K, and schizophyllan. However, there is currently insufficient clinical data to support the effectiveness and safety of these extracts in humans. While some mushrooms or their extracts may be used for medicinal purposes, regulatory authorities for drugs, such as the US Food and Drug Administration, consider mushroom consumption to be a nutritional supplement. This is because it lacks official government approval and is not commonly used as a pharmaceutical drug in clinical settings [8, 73, 102] (Figure 6.13).

**FIGURE 6.13** *Ganoderma lingzhi.*

*Source*: Image Number 574883 at Mushroom Observer, a source for mycological images. https://creativecommons.org/licenses/by-sa/3.0/

### 6.9.4 OTHER USES

You can color wool and other natural fibers using mushrooms. Mushroom dyes can produce all colors of the rainbow because their organic molecules, known as chromophores, create intense and vivid hues. Before synthetic dyes were developed, many textile dyes were derived from mushrooms [77] (Figure 6.14).

**FIGURE 6.14** A tinder fungus, *Fomes fomentarius*.

*Source*: Photo by George Chernilevsky. Publilc domain.

Some fungi, especially varieties of *polypores,* have been used to ignite fires. New biological remediation strategies, such as those that use filtering technology and mycorrhizae to boost plant growth, are being developed in part thanks to mushrooms and other fungi (e.g., utilizing fungi to reduce bacterial numbers in polluted water). A genetic engineering study is now being conducted to improve the properties of mushrooms for use in medicine and enhance their nutritional value, among other things [16, 115].

### 6.9.5 *PSYCHOACTIVE MUSHROOMS*

Many indigenous medical practices have used hallucinogenic mushrooms for a very long time in different civilizations all over the world [14]. They have been utilized as sacraments in rituals intended to promote physical and mental healing, as well as visionary experiences. One of these rituals is the velada ceremony [41]. The shaman or curandera, who is a priest-healer, uses mushrooms in their traditional manner [14, 25] (Figure 6.15).

**FIGURE 6.15** *Psilocybe zapotecorum*, a hallucinogenic mushroom.

*Source*: Photo by Alan Rockefeller. https://creativecommons.org/licenses/by-sa/4.0/

Mushrooms that contain psilocybin, popularly known as psychedelic mushrooms, have psychedelic qualities [26, 84]. Commonly referred to as "magic mushrooms" or "shrooms," they are freely sold in upscale stores across the world or are illegal in nations where their sale is prohibited [48].

It has been claimed that psilocybin mushrooms can lead to profound, life-altering discoveries that are frequently regarded as mystical experiences [26]. These statements, as well as the enduring impacts of such artificially generated spiritual experiences, have been validated by recent scientific research [74, 79, 108] (Figure 6.16).

**FIGURE 6.16** Native regions of psychoactive mushroom of genus *Psilocybe*.

*Source*: Photo by Pastelitodepapa. https://creativecommons.org/publicdomain/zero/1.0/deed.en

Researchers are investigating the potential use of natural psilocybin, a compound found in *Psilocybe cubensis* and other mushroom species, for the treatment of psychological disorders, including obsessive-compulsive disorder [5]. It has been suggested that even small doses can alleviate cluster and migraine headaches. According to a double-blind study conducted at Johns Hopkins Hospital, psychedelic mushrooms can induce deeply meaningful and spiritually significant experiences for users. One-third of the participants in the survey reported that using psychedelic mushrooms was the most important spiritual experience of their lives. More than two-thirds of the participants listed it as one of the top five significant and spiritually impactful events. However, one-third of the participants reported experiencing severe anxiety, which subsided after a short period of time [32]. Psilocybin mushrooms have also shown effectiveness in treating addiction, particularly to alcohol and nicotine [86].

The hallucinogenic compound muscimol is found in certain species of the *Amanita* genus, notably *A. pantherina* and *A. muscaria* [70]. The muscimol-containing cytotoxic drugs in the *Amanitas* do not contain amatoxins or phallotoxins, making them non-hepatotoxic. However, if not handled properly, the presence of ibotenic acid can make them neurotoxic without being lethal. Amanita intoxication is similar to Z-drugs in terms of its effects on the central nervous system, inducing sedation and a calming, mesmerizing effect, as well as detachment and restlessness in high doses [81, 97].

## 6.10 HISTORY OF USING MUSHROOM AS ANTIDEPRESSION MEDICINE

### 6.10.1 *EARLY HISTORY*

According to prehistoric rock art at Villar del Humo in Spain, *Psilocybe hispanica* was used in religious ceremonies 6,000 years ago [3]. Native Americans in Mesoamerica have utilized hallucinogenic species of the *Psilocybe* genus for spiritual empathy, prediction, and healing since pre-Columbian times [14]. In Guatemala, mushroom stones and patterns have been discovered. A sculpture depicting a mushroom resembling *Psilocybe mexicana,* dating back to around 200 CE, was found in a pyramid and compartment monument in the Colima state of western Mexico. According to legend, the Aztecs served tenancatl (a combination of the Nahuatl words tetl, meaning "god, sacred," and nancatl, meaning "mushroom") at the coronation of Moctezuma II in 1502. Psilocybin mushrooms were known to the Aztecs and Mazatecs as genius mushrooms, divinatory mushrooms, and magnificent mushrooms when translated into English. After Hernán Cortés' voyage, Bernardino de Sahagún visited Central America and observed how the Aztecs used *teonanácatl* in rituals [46].

The Catholic crusaders launched a social attack on the Aztecs, labeling them as idolaters. Consequently, the use of hallucinogenic plants and mushrooms, along with other pagan practices, quickly became illegal. The Spanish claimed that the fungus allowed the Aztecs to communicate with demons. Despite its past, *teonanácatl* is still spoken of in certain rural areas [34].

### 6.10.2 MODERN HISTORY

The earliest reference to hallucinogenic mushrooms in European medical works appeared in 1799 in the London Medical and Physical Journal. It tells the story of a man who fed his family some *Psilocybe semilanceata* mushrooms he had picked for breakfast in London's Green Park. After consuming them, the youngest child experienced uncontrollable bouts of giggling that neither his father nor mother could contain [11].

The first European Americans to enthusiastically participate in a native mushroom ceremony were R. Gordon Wasson and Valentina Pavlovna Wasson in 1955. The Wassons made significant efforts to share their experience, even writing an article about it for Life on May 13, 1957 [111]. In 1956, Roger Heim named the psychedelic fungus that the Wassons brought back from Mexico as Psilocybe, and in 1958, Albert Hofmann discovered the active substances psilocybin and psilocin in these mushrooms [44, 45].

Timothy Leary read the Wassons' story in Life and then traveled to Mexico to try psilocybin mushrooms. When he returned to Harvard in 1960, he and Richard Alpert established the Harvard Psilocybin Project, which aimed to study the psychological and theological effects of psilocybin and other psychedelic compounds. In the 1960s, Alpert and Leary planned to examine the impact of psilocybin on recidivism among prisoners [23]. When the patients were reevaluated six months later, the recidivism rate had dropped to 40%, which was lower than their predictions. This, along with another experiment involving graduate divinity students being administered psilocybin, sparked discussions. After being dismissed from Harvard in 1963, Leary and Alpert focused their efforts on promoting the psychedelic experience within the emerging hippie subculture [23].

The Terence McKenna, Wassons, Leary, Robert Anton Wilson, and many more made entheogens more well-known, which led to a rise in psilocybin mushroom usage all over the world. Several psilocybin species of mushrooms from Europe, North America, as well as Asia had been identified and extensively collected by the early 1970s. There were other books released that described how to cultivate vast quantities of *Psilocybe cubensis*. Psilocybin mushrooms have become one of the most extensively used psychedelic substances due to their availability from both wild and farmed sources [111].

Currently, psilocybin mushroom use has been documented among certain Nahua, Mixtec, Mixe, Mazatec, Zapotec, and other communities stretching from central Mexico to Oaxaca [34]. María Sabina, a prominent figure in Mexican mushroom usage, used medicinal fungi like *Psilocybe mexicana* in their treatment [72].

## 6.11 CHEMICAL COMPOSITION OF MAGIC MUSHROOMS

The chemical composition of magic mushroom species and genera varies greatly from one another [34]. Its main active ingredient is psilocybin [26], which produces the psychedelic experience when converted into psilocin. Additionally, the presence of the compounds psilocin, norpsilocin, baeocystin, norbaeocystin, and aeruginascin may alter the effects of psychedelic mushrooms [107]. Another type of psychedelic mushroom, *Panaeolus subbalteatus,* contains the highest amount of psilocybin compared to other mycelium [99]. Some mushrooms have been found to inhibit the enzyme monoamine oxidase, responsible for breaking down tryptamine alkaloids, by forming beta-carbolines. Beta-carbolines are present in *Cyclocybe, Hygrophorus*, and *Psilocybe*. Psilocybe species have been shown to contain harmine and a variety of additional l-tryptophan-derived β-carbolines [9, 21].

## 6.12 EFFECTS OF PSILOCYBIN MUSHROOMS

Psilocybin and psilocin are the active ingredients in psilocybin mushrooms. The liver degrades psilocybin through a mechanism known as dephosphorylation after consumption. The resulting molecule, psilocin, is what produces the psychedelic effects [62]. It is difficult to abuse psilocybin and psilocin since they are also delivered in a brief period of time, which reduces their advantages [101]. There is no proof that psilocybin mushrooms cause physical or mental dependence (addiction) [108]. Usually starting 20 minutes after intake, the psychedelic effects might boost for a maximum of six hours. Possible adverse effects include exhilaration, vomiting, nausea, lethargy or muscular relaxation, fatigue, and dizziness [108].

Like other psychedelic medications, psychedelic mushrooms have subjective effects that can differ considerably from user to user. Based on the dose, manner of manufacturing, and human digestion, the mind-altering effects of psilocybin-containing mushrooms can last anywhere from three to eight hours. It is typical to refer to the first 3–4 hours after intake, when the consumer encounters more intense sights and perceptual abnormalities, as the "peak." Due to psilocybin's capacity to alter temporal consciousness, the consumer might believe that the effects continue much longer than they actually do [114].

### 6.12.1 SENSORY EFFECTS

Multisensory hallucination, psychological symptoms, and other sensory consequences changed spatial or temporal perceptions [88]. Effects may take up to two hours to become visible, however multisensory, and physical sensations may change 30 to 60 minutes after ingestion. Color enhancement and contrast, strange illumination phenomena, improved visual insight, exteriors that appear to ripple, shimmer, or breathe, intricate open and closed eye illustrations of form constants or images, substances that warp, morph, or change solid colors, a sense of melting into the situation, and trails behindhand touching objects are some examples of these changes in perception. The clarity of sounds may increase; for example, music may develop a rich feeling of cadence and depth [88]. Occasionally patients have synesthesia, during which they perceive something after hearing it, such as a colored image [19].

### 6.12.2 EMOTIONAL EFFECTS

The experience, or 'trip,' with psychedelics such as Lysergic acid diethylamide, heavily depends on the environment and circumstances [19]. Prevalent adverse reactions include laughter, loss of focus, muscle relaxation (including dilated pupils), often occurring together [19]. While a cozy and familiar setting can contribute to a pleasurable experience, uncomfortable circumstances may result in a miserable trip. Psychedelics intensify emotions, so if a person is anxious before a trip, they are likely to feel even more anxious during one. Many individuals choose to consume psychedelics with peer groups or someone who has previous "tripping" experience. The mental consequences of psychedelics usage include illusions and the inability to distinguish between fiction and reality. Additionally, panic attacks and insanity are possible, especially if a large amount is consumed. Those who use psilocybin mushrooms face the risk of toxic poisoning if one of the many types of deadly mushrooms is mistaken for psychedelic mushrooms, in addition to the risks associated with psilocybin ingestion [19].

## 6.13 DOSAGE

The dosage of psilocybin-containing mushrooms is determined by the concentration of psilocybin and psilocin in the mushrooms, which can vary greatly between and within species. Typically, the concentration is approximately 0.5 to 2.0% of the mushroom's dry weight (DW). For *Psilocybe cubensis,* doses usually range from 1.0 to 2.5 g, with 2.5 to 5.0 g of dry mushroom material considered a strong dosage. More than 5 g of dry fungus is often referred to as a "high dose," with 5.0 g being called a "heroic dose" [51].

Not only do psilocybin components vary between species, but mushrooms within a single species, subspecies, or variation can also differ in terms of the amount of active psilocybin they contain [7]. The highest concentration of psilocybin, ranging between 0.23% and 0.90%, is found in the dry caps of the mushroom *Psilocybe samuiensis*. The mycelium makes up 0.24 to 0.32% [27].

## 6.14 HOW PSILOCYBIN MUSHROOMS WORK?

Our brains rely on vast networks of nerve cells that communicate with one another via chemical messengers known as neurotransmitters. Astrocytes, which are brain cells, produce hormones that regulate the brain's blood flow, which in turn controls activity levels. These two systems work together to keep our ideas, actions, and emotions in line [66, 76, 112].

Psilocybin, which is present in mushrooms, momentarily disrupts the brain's functions, producing a form of "reset" that results in greater nerve propagation and interaction, stabilized neurotransmitter levels, and boosted blood circulation to brain regions with low activity [82]. The use of mushrooms for curing depression has both pros and cons. Despite modern advancements in human health, a complete understanding of the pros and cons of using psilocybin has not yet been developed in less than 10 years. Psilocybin extracted from mushrooms can provide relief from all types of anxiety accompanied by depression [33, 40]. In the United States, access to Psilocybin is very limited, and most types of mushroom drugs are illegal except those accepted by the FDA. Psilocybin possesses no side effects except for hallucinations, which typically disappear a few hours after administration, although they may be unclear [33, 40].

## 6.15 CLINICAL EXPERIENCE

The use of psilocybin for treating depression is vastly different from recreational use. A highly purified form of psilocybin has been utilized in various clinics for mental health treatment. It is administered to patients in a controlled environment under the supervision of medical professionals to ensure both effectiveness and safety [69].

## 6.16 TREATMENTS FOR DEPRESSION

Transcranial magnetic stimulation (TMS), which is a quick, noninvasive, and painless technique, has been employed to treat mental illness since 2008. A strong and detailed magnetic field is administered to a part of the brain that controls mood during TMS. Through TMS, many patients experience no side effects and long-lasting alleviation of depression [71].

## 6.17 SUMMARY

The Himalayan region is abundant in vegetation and wildlife. The diverse locations in the Himalayas are known to host a wide variety of mushrooms. It is well-documented that many cultures utilize wild mushrooms for both culinary and medicinal purposes. Mushrooms contain several natural antibiotics. The immunomodulatory properties of glucans found in the cell wall are widely recognized, and secondary metabolites have demonstrated effectiveness against bacteria and viruses. Mushroom mycelia exudates have been successful in eradicating protozoans. Psilocybin-containing mushrooms, commonly referred to as psychedelic mushrooms, are known for their psychoactive effects. Psychedelic mushrooms have been used to treat conditions such as sadness. Psilocybin, a substance found in mushrooms, temporarily halts brain activity and resets it, resulting in improved nerve transmission, communication, and reduced anxiety levels. As with any medical intervention, there are pros and cons to using psilocybin for depression treatment. Further research is needed to fully understand the potential and availability of mushroom species, as well as to optimize manufacturing techniques to enhance the quality and quantity of antidepressant compounds present in mushrooms. Developing mushroom varieties with higher psilocybin yields, establishing a rapid propagation protocol for commercially sought-after mushroom species, and cultivating mushrooms for the production of antidepressants and other industrial applications all require the utilization of biotechnological and modern approaches.

## KEYWORDS


- **antidepression**
- **depression**
- **ethno-mycology**
- **hallucinogenic**
- **Himalayan**
- **mushrooms**
- **psilocybin**
- **psychedelic mushrooms**
- **psychedelic qualities**
- **psychoactive mushroom**


## REFERENCES


1. Dickie, I. A., & Reich, P. B., (2005). Ectomycorrhizal fungal communities at forest edges. *Journal of Ecology, 93*, 244–255.
2. Abdelfattah, A., Wisniewski, M., Li Destri, N. M. G., Cacciola, S. O., & Schena, L., (2016). Metagenomic analysis of fungal diversity on strawberry plants and the effect of management practices on the fungal community structure of aerial organs. *PloS One, 11,* e0160470.
3. Akers, B. P., Ruiz, J. F., Piper, A., & Ruck, C. A., (2011). A prehistoric mural in Spain depicting neurotropic *Psilocybe mushrooms*? 1. *Economic Botany, 65*, 121–128.
4. Baker, T., (1990). The word 'toadstool' in Britain. *Mycologist, 4*, 25–29.
5. Bauer, B. E., (2020). *Aeruginascin Identified in Psilocybe Cubensis Magic Mushrooms.* https://psychedelicreview.com/aeruginascin-identified-in-psilocybe-cubensis-magic-mushrooms/ (accessed on 2 March 2024).
6. Beug, M., Bessette, A. E., & Bessette, A. R., (2014). *Ascomycete Fungi of North America: A Mushroom Reference Guide*. University of Texas Press.
7. Bigwood, J., & Beug, M. W., (1982). Variation of psilocybin and psilocin levels with repeated flushes (harvests) of mature sporocarps of *Psilocybe cubensis* (Earle) singer. *Journal of Ethnopharmacology, 5*, 287–291.
8. Blagodatski, A., Yatsunskaya, M., Mikhailova, V., Tiasto, V., Kagansky, A., & Katanaev, V. L., (2018). Medicinal mushrooms as an attractive new source of natural compounds for future cancer therapy. *Oncotarget, 9*, 29259.

9. Blei, F., Dörner, S., Fricke, J., Baldeweg, F., Trottmann, F., Komor, A., Meyer, F., et al., (2020). Simultaneous production of psilocybin and a cocktail of β-carboline monoamine oxidase inhibitors in "magic" mushrooms. *Chemistry–A European Journal, 26*, 729–734.
10. Boa, E. R., (2004). *Wild Edible Fungi: A Global Overview of Their Use and Importance to People, 17*, 1–70.
11. Brande, E., (1800). Mr. E. Brande, on a poisonous species of agaric. *The Medical and Physical Journal, 3*, 41.
12. Burdsall, H., & Volk, T. J., (2008). Armillaria solidipes, an older name for the fungus called *Armillaria ostoyae*. *North American Fungi, 3*, 261–267.
13. Cardwell, G., Bornman, J. F., James, A. P., & Black, L. J., (2018). A review of mushrooms as a potential source of dietary vitamin D. *Nutrients, 10*, 1498.
14. Carod-Artal, F. J., (2015). Hallucinogenic drugs in pre-Columbian Mesoamerican cultures. *Neurología (English Edition), 30*, 42–49.
15. Chandra M. R., (2011). *Estimation of Enzymatic Activities of Different Species of Mushrooms*. Doctoral Dissertation.
16. Chang, S. T., (2006). The world mushroom industry: Trends and technological development. *International Journal of Medicinal Mushrooms, 8*.
17. Chelliah, B. J., Kalaiarasi, S., Anand, A., Janakiram, G., Rathi, B., & Warrier, K., (2018). Classification of mushrooms using supervised learning models. *International Journal of Emerging Technologies in Engineering Research (IJETER), 6*.
18. Chen, W. Y., Yang, T., Yang, J., Qiu, Z. C., Ding, X. Y., Wang, Y. H., & Wang, Y. H., (2021). Wild plants used by the Lhoba people in Douyu Village, characterized by high mountains and valleys, in southeastern Tibet, China. *Journal of Ethnobiology and Ethnomedicine, 17*, 1–21.
19. Cole, S. M., (2006). *New Research on Street Drugs*. Nova Publishers.
20. Cotter, T., (2014). *Organic Mushroom Farming and Mycoremediation: Simple to Advanced and Experimental Techniques for Indoor and Outdoor Cultivation*. Chelsea Green Publishing.
21. Matzopoulos, R., Morlock, R., Morlock, A., Lerer, B., & Lerer, L., (2022). Psychedelic mushrooms in the USA: Knowledge, patterns of use, and association with health outcomes. *Frontiers in Psychiatry, 12*, 780696.
22. Dargan, J., (2016). *My Dates with Perithecial Fungi, 47*, 27–34.
23. Doblin, R., (1998). Dr. Leary's concord prison experiment: A 34-year follow-up study. *Journal of Psychoactive Drugs, 30*, 419–426.
24. Ekanayaka, A., Ariyawansa, H., Hyde, K., Jones, E., Daranagama, D., Phillips, A., Hongsanan, S., et al., (2017). Discomycetes: The apothecial representatives of the phylum *Ascomycota*. *Fungal Diversity, 87*, 237–298.
25. Fokunang, E. T., Annih, M. G., Abongwa, L. E., Bih, M. E., Vanessa, T. M., Fomnboh, D. J., & Fokunang, C., (2022). *Medicinal Mushroom of Potential Pharmaceutical Toxic Importance: Contribution in Phytotherapy*.
26. Freye, E., (2009). The mushroom psilocybin with psychedelic properties. *Pharmacology and Abuse of Cocaine, Amphetamines, Ecstasy, and Related Designer Drugs* (pp. 221–223). Springer.
27. Gartz, J., Allen, J. W., & Merlin, M. D., (1994). Ethnomycology, biochemistry, and cultivation of *Psilocybe samuiensis* Guzman, Bandala and Allen, a new psychoactive fungus from Koh Samui, Thailand. *Journal of Ethnopharmacology, 43*, 73–80.

28. Genders, R., (1982). *Mushroom Growing for Everyone.* Faber and Faber.
29. Ghorbani, A., Langenberger, G., Liu, J. X., Wehner, S., & Sauerborn, J., (2012). Diversity of medicinal and food plants as non-timber forest products in Naban River Watershed National Nature Reserve (China): Implications for livelihood improvement and biodiversity conservation. *Economic Botany, 66*, 178–191.
30. Gillman, L., (1994). Identification of common poisonous mushrooms. *Handbook of Mushroom Poisoning: Diagnosis and Treatment, 97.*
31. Giusti, A., Ricci, E., Gasperetti, L., Galgani, M., Polidori, L., Verdigi, F., Narducci, R., & Armani, A., (2020). Molecular identification of mushroom species in Italy: An ongoing project aimed at reinforcing the control measures of an increasingly appreciated sustainable food. *Sustainability, 13*, 238.
32. Griffiths, R. R., Richards, W. A., McCann, U., & Jesse, R., (2006). Psilocybin can occasion mystical-type experiences having substantial and sustained personal meaning and spiritual significance. *Psychopharmacology, 187*, 268–283.
33. Guo, J. Y., Han, C. C., & Liu, Y. M., (2010). A contemporary treatment approach to both diabetes and depression by cordyceps sinensis, rich in vanadium. *Evidence-Based Complementary and Alternative Medicine, 7*, 387–389.
34. Guzmán, G., (2008). Hallucinogenic mushrooms in Mexico: An overview. *Economic Botany, 62*, 404–412.
35. Hajeb, P., & Jinap, S., (2015). Umami taste components and their sources in Asian foods. *Critical Reviews in Food Science and Nutrition, 55*, 778–791.
36. Hajslová, J., & Schulzova, V., (2008). Mushroom toxins. *Bioactive Compounds in Foods*, 111–133.
37. Hall, A. H., Spoerke, D. G., & Rumack, B. H., (1987). Mushroom poisoning: Identification, diagnosis, and treatment. *Pediatrics in Review, 8*, 291–298.
38. Halme, P., & Kotiaho, J. S., (2012). The importance of timing and number of surveys in fungal biodiversity research. *Biodiversity and Conservation, 21*, 205–219.
39. Hanif, M., Khalid, A. N., Sarwar, S., & Yousaf, N., (2022). Ectomycorrhizal status of *Pinus wallichiana* (blue pine) growing in Himalayan moist temperate Forests of Pakistan. *Pak. J. Bot., 54*, 275–283.
40. He, M. Q., Wang, M. Q., Chen, Z. H., Deng, W. Q., Li, T. H., Alfredo, V., Jeewon, R., et al., (2022). Potential benefits and harms: A review of poisonous mushrooms in the world. *Fungal Biology Reviews.*
41. Hernández-Ávila, I., (1996). Mediations of the spirit: Native American religious traditions and the ethics of representation. *American Indian Quarterly, 20*, 329–352.
42. Hibbett, D. S., Grimaldi, D., & Donoghue, M. J., (1997). Fossil mushrooms from Miocene and Cretaceous ambers and the evolution of Homobasidiomycetes. *American Journal of Botany, 84*, 981–991.
43. Hladký, J., (2013). Terminology. *The Czech and the English Names of Mushrooms* (pp. 13–25). https://digilib.phil.muni.cz/sites/default/files/pdf/122814.pdf (accessed on 2 March 2024).
44. Hofmann, A., Frey, A., Ott, H., Petrzilka, T., & Troxler, F., (1958). Konstitutionsaufklärung und synthese von psilocybin. *Experientia, 14*, 397–399.
45. Hofmann, A., Heim, R., Brack, A., & Kobel, H., (1958). Psilocybin, ein psychotroper Wirkstoff aus dem mexikanischen RauschpilzPsilocybe mexicana Heim. *Experientia, 14*, 107–109.

46. Hofmann, A., (1980). *The Mexican Relatives of LSD.* LSD: My Problem Child. New York: McGraw-Hill.
47. Hossain, M. I., (2016). *Effect of Different Sawdust and Effect of Different Substrate on the Growth, Yield and Proximate Composition of Shiitake Mushroom (Lentinus edudes).* Dept. of Biochemistry.
48. Jay, M., (2010). *High Society: The Central Role of Mind-Altering Drugs in History, Science, and Culture.* Simon and Schuster.
49. Jayachandran, M., Xiao, J., & Xu, B., (2017). A critical review on health promoting benefits of edible mushrooms through gut microbiota. *International Journal of Molecular Sciences, 18,* 1934.
50. Jayasiri, S. C., Hyde, K. D., Ariyawansa, H. A., Bhat, J., Buyck, B., Cai, L., Dai, Y. C., et al., (2015). The faces of fungi database: Fungal names linked with morphology, phylogeny, and human impacts. *Fungal Diversity, 74,* 3–18.
51. Jesso, J. W., (2013). *Decomposing the Shadow: Lessons from the Psilocybin Mushroom.* Souls Lantern Publishing.
52. Jinlong, L., Renhua, Z., & Qiaoyun, Z., (2012). Traditional forest knowledge of the Yi people confronting policy reform and social changes in Yunnan province of China. *Forest Policy and Economics, 22,* 9–17.
53. Jo, W. S., Hossain, M. A., & Park, S. C., (2014). Toxicological profiles of poisonous, edible, and medicinal mushrooms. *Mycobiology, 42,* 215–220.
54. Kotowski, M. A., Molnár, Z., & Łuczaj, Ł., (2021). Fungal ethnoecology: Observed habitat preferences and the perception of changes in fungal abundance by mushroom collectors in Poland. *Journal of Ethnobiology and Ethnomedicine, 17,* 1–23.
55. Koutrotsios, G., Larou, E., Mountzouris, K. C., & Zervakis, G. I., (2016). Detoxification of olive mill wastewater and bioconversion of olive crop residues into high-value-added biomass by the choice edible mushroom *Hericium erinaceus. Applied Biochemistry and Biotechnology, 180,* 195–209.
56. Kuees, U., & Navarro-Gonzalez, M., (2015). How do agaricomycetes shape their fruiting bodies? 1. morphological aspects of development. *Fungal Biology Reviews, 29,* 63–97.
57. Kües, U., & Liu, Y., (2000). Fruiting body production in basidiomycetes. *Applied Microbiology and Biotechnology, 54,* 141–152.
58. Lechner, B. E., Wright, J., & Albertó, E., (2005). The genus Pleurotus in Argentina: Mating tests. *SYDOWIA-HORN, 57,* 233.
59. Lee, J. S., Lim, M. O., Cho, K. Y., Cho, J. H., Chang, S. Y., & Nam, D. H., (2006). Identification of medicinal mushroom species based on nuclear large subunit rDNA sequences. *Journal of Microbiology, 44,* 29–34.
60. Li, C., & Xu, S., (2022). Edible mushroom industry in China: Current state and perspectives. *Applied Microbiology and Biotechnology,* 1–7.
61. Lima, A. D., Fortes, R. C., Novaes, M. G., & Percário, S., (2012). Poisonous mushrooms; a review of the most common intoxications. *Nutricion Hospitalaria, 27,* 402–408.
62. Ling, S., Ceban, F., Lui, L. M., Lee, Y., Teopiz, K. M., Rodrigues, N. B., Lipsitz, O., et al., (2021). Molecular mechanisms of psilocybin and implications for the treatment of depression. *CNS Drugs,* 1–14.

63. Litten, W., (1975). The most poisonous mushrooms. *Scientific American, 232*, 90–101.
64. Liu, D., Cheng, H., Bussmann, R. W., Guo, Z., Liu, B., & Long, C., (2018). An ethnobotanical survey of edible fungi in Chuxiong City, Yunnan, China. *Journal of Ethnobiology and Ethnomedicine, 14*, 1–10.
65. Luther, B., (2010). *Spore Prints* (pp. 1–8). https://www.psms.org/sporeprints/SP462.pdf (accessed on 2 March 2024).
66. Lutkajtis, A., (2020). Lost saints: Desacralization, spiritual abuse and magic mushrooms. *Fieldwork in Religion, 14*, 118–139.
67. Ma, G., Yang, W., Zhao, L., Pei, F., Fang, D., & Hu, Q., (2018). A critical review on the health promoting effects of mushrooms nutraceuticals. *Food Science and Human Wellness, 7*, 125–133.
68. Mamiro, D. P., & Royse, D. J., (2008). The influence of spawn type and strain on yield, size, and mushroom solids content of *Agaricus bisporus* produced on non-composted and spent mushroom compost. *Bioresource Technology, 99*, 3205–3212.
69. Meyer, J. S., (2013). 3, 4-methylenedioxymethamphetamine (MDMA): Current perspectives. *Substance Abuse and Rehabilitation, 4*, 83.
70. Michelot, D., & Melendez-Howell, L. M., (2003). Amanita muscaria: Chemistry, biology, toxicology, and ethnomycology. *Mycological Research, 107*, 131–146.
71. Mishra, B. R., Sarkar, S., Praharaj, S. K., Mehta, V. S., Diwedi, S., & Nizamie, S. H., (2011). Repetitive transcranial magnetic stimulation in psychiatry. *Annals of Indian Academy of Neurology, 14*, 245.
72. Monaghan, J. D., & Edmonson, B. W., (2000). *Supplement to the Handbook of Middle American Indians: Ethnology* (Vol. 6). University of Texas Press.
73. Money, N. P., (2016). Are mushrooms medicinal? *Fungal biology, 120*, 449–453.
74. Nkadimeng, S. M., Nabatanzi, A., Steinmann, C. M., & Eloff, J. N., (2020). Phytochemical, cytotoxicity, antioxidant, and anti-inflammatory effects of psilocybe natalensis magic mushroom. *Plants, 9*, 1127.
75. Oei, P., & Nieuwenhuijzen, B. V., (2005). *Small-Scale Mushroom Cultivation*. Agromisa/CTA.
76. Oss, O., & Oeric, O., (1992). *Psilocybin: Magic Mushroom Grower's Guide: A Handbook for Psilocybin Enthusiasts*. Rosenthal.
77. Özdemir, H., & Bozok, F., (2020). Dyeing of wool yarn with natural dyes of *Lactarius deliciosus* and *L. sanguifluus* from Turkey. *Textile and Apparel, 30*, 262–269.
78. Parveen, A., Khataniar, L., Goswami, G., Hazarika, D. J., Das, P., Gautom, T., Barooah, M., & Boro, R. C., (2017). A study on the diversity and habitat specificity of macrofungi of Assam, India. *International Journal of Current Microbiology and Applied Sciences, 6*, 275–297.
79. Peden, N. R., Macaulay, K., Bissett, A. F., Crooks, J., & Pelosi, A., (1981). Clinical toxicology of 'magic mushroom' ingestion. *Postgraduate Medical Journal, 57*, 543–545.
80. Phillips, J. M., Ooi, S. L., & Pak, S. C., (2022). Health-promoting properties of medicinal mushrooms and their bioactive compounds for the COVID-19 era—An appraisal: Do the pro-health claims measure up? *Molecules, 27*, 2302.
81. Phipps, A. G., (2000). *Japanese use of Beni-Tengu-Dake (Amanita Muscaria) and the Efficacy of Traditional Detoxification Methods* (pp. 8–80). https://digitalcommons.fiu.edu/cgi/viewcontent.cgi?article=6031&context=etd (accessed on 2 March 2024).

82. Powell, S. G., (2011). *The Psilocybin Solution: The Role of Sacred Mushrooms in the Quest for Meaning.* Simon and Schuster.
83. Rajarathnam, S., Bano, Z., & Miles, P. G., (1987). Pleurotus mushrooms. Part I A. Morphology, life cycle, taxonomy, breeding, and cultivation. *Critical Reviews in Food Science & Nutrition, 26*, 157–223.
84. Ramírez-Cruz, V., Guzmán, G., Villalobos-Arámbula, A. R., Rodríguez, A., Matheny, P. B., Sánchez-García, M., & Guzmán-Dávalos, L., (2013). Phylogenetic inference and trait evolution of the psychedelic mushroom genus *Psilocybe* sensu lato (*Agaricales*). *Botany, 91*, 573–591.
85. Rathore, H., Prasad, S., & Sharma, S., (2017). Mushroom nutraceuticals for improved nutrition and better human health: A review. *Pharma Nutrition, 5*, 35–46.
86. Roberts, C. A., Osborne-Miller, I., Cole, J., Gage, S. H., & Christiansen, P., (2020). Perceived harm, motivations for use and subjective experiences of recreational psychedelic 'magic' mushroom use. *Journal of Psychopharmacology, 34*, 999–1007.
87. Rosmiza, M., Davies, W., CR, R. A., Jabil, M., & Mazdi, M., (2016). Prospects for increasing commercial mushroom production in Malaysia: Challenges and opportunities. *Mediterranean Journal of Social Sciences, 7*, 406.
88. Schultes, R. E., & Smith, E. W., (1976). *Hallucinogenic Plants.* Golden Press New York.
89. Scimeca, J. A., (1997). Naturally occurring orally active dietary carcinogens. In: Massaro, E. J., (ed.), *Handbook of Humantoxicology* (pp. 409–466). Boka Raton, New York, USA.
90. Semwal, K., Stephenson, S., Bhatt, V., & Bhatt, R., (2014). Edible mushrooms of the Northwestern Himalaya, India: A study of indigenous knowledge, distribution. and diversity. *Mycosphere. 5*, 440–461.
91. Shepard, G. H., Arora, D., & Lampman, A., (2008). The grace of the flood: Classification and use of wild mushrooms among the highland maya of Chiapas. *Economic Botany, 62*, 437–470.
92. Sher, H., Aldosari, A., & Bussmann, R. W., (2015). Morels of Palas Valley, Pakistan: A potential source for generating income and improving livelihoods of mountain communities. *Economic Botany, 69*, 345–359.
93. Siachoono, S. M., Shakachite, O., Muyenga, A. M., & Bwalya, J., (2016). Underground treasure: A preliminary inquiry into the ecology and distribution of Zambian truffles. *International Journal of Biology, 8*, 1–8.
94. Simon, R. R., Phillips, K. M., Horst, R. L., & Munro, I. C., (2011). Vitamin D mushrooms: Comparison of the composition of button mushrooms (*Agaricus bisporus*) treated postharvest with UVB light or sunlight. *Journal of Agricultural and Food Chemistry, 59*, 8724–8732.
95. Sitotaw, R., Lulekal, E., & Abate, D., (2020). Ethnomycological study of edible and medicinal mushrooms in Menge District, Asossa zone, Benshangul Gumuz region, Ethiopia. *Journal of Ethnobiology and Ethnomedicine, 16*, 1–14.
96. Smith, A. H., Weber, N. S., & Weber, N. S., (1980). *The Mushroom Hunter's Field Guide.* University of Michigan Press.
97. Spoerke, D. G., (1994). Gastrointestinal irritant mushrooms. *Handbook of Mushroom Poisoning—Diagnosis and Treatment* (pp. 347–366). CRC Press, Boca Raton.

98. Spoerke, D. G., (2019). Fungi in folk medicine and society. *Foodborne Disease Handbook* (pp. 781–802). CRC Press.
99. Stijve, T., & Kuyper, T. W., (1985). Occurrence of psilocybin in various higher fungi from several European countries. *Planta Medica, 51*, 385–387.
100. Strausfeld, N. J., Wolff, G. H., & Sayre, M. E., (2020). Mushroom body evolution demonstrates homology and divergence across *Pancrustacea. Elife, 9.*
101. Studerus, E., Kometer, M., Hasler, F., & Vollenweider, F. X., (2011). Acute, subacute, and long-term subjective effects of psilocybin in healthy humans: A pooled analysis of experimental studies. *Journal of Psychopharmacology, 25*, 1434–1452.
102. Sullivan, R., Smith, J. E., & Rowan, N. J., (2006). Medicinal mushrooms and cancer therapy: Translating a traditional practice into Western medicine. *Perspectives in Biology and Medicine, 49*, 159–170.
103. Tang, C., Hoo, P. C. X., Tan, L. T. H., Pusparajah, P., Khan, T. M., Lee, L. H., Goh, B. H., & Chan, K. G., (2016). Golden needle mushroom: A culinary medicine with evidenced-based biological activities and health promoting properties. *Frontiers in Pharmacology, 7*, 474.
104. Tegzes, J. H., & Puschner, B., (2002). Toxic mushrooms. *Veterinary Clinics: Small Animal Practice, 32*, 397–407.
105. Treschow, C., (1944). *Nutrition of the Cultivated Mushroom* (p. 190). https://www.cabidigitallibrary.org/doi/full/10.5555/19450300754 (accessed on 2 March 2024).
106. Tsing, A., (2012). Unruly edges: Mushrooms as companion species for donna Haraway. *Environmental Humanities, 1*, 141–154.
107. Tsujikawa, K., Kanamori, T., Iwata, Y., Ohmae, Y., Sugita, R., Inoue, H., & Kishi, T., (2003). Morphological and chemical analysis of magic mushrooms in Japan. *Forensic Science International, 138*, 85–90.
108. Van, A. J., Opperhuizen, A., & Van, D. B. W., (2011). Harm potential of magic mushroom use: A review. *Regulatory Toxicology and Pharmacology, 59*, 423–429.
109. Venturella, G., & Zervakis, G., (2006). *Pleurotus nebrodensis*. The IUCN Red List of Threatened Species.
110. Wasser, S., (2002). Medicinal mushrooms as a source of antitumor and immunomodulating polysaccharides. *Applied Microbiology and Biotechnology, 60*, 258–274.
111. Wasson, R. G., (1957). Seeking the magic mushroom. *Life, 42*, 100–120.
112. Watling, R., (1983). Hallucinogenic mushrooms. *Journal of the Forensic Science Society, 23*, 53–66.
113. White, H. J., León-Sánchez, L., Burton, V. J., Cameron, E. K., Caruso, T., Cunha, L., Dirilgen, T., et al., (2020). Methods and approaches to advance soil macroecology. *Global Ecology and Biogeography, 29*, 1674–1690.
114. Wittmann, M., Carter, O., Hasler, F., Cahn, B. R., Grimberg, U., Spring, P., Hell, D., et al., (2007). Effects of psilocybin on time perception and temporal control of behavior in humans. *Journal of Psychopharmacology, 21*, 50–64.
115. Zjawiony, J. K., (2004). Biologically active compounds from *Aphyllophorales* (polypore) fungi. *Journal of Natural Products, 67*, 300–310.

# PART III

# Antidepressant Agents from Himalayan Beverages, Spices, and Microalgae

## CHAPTER 7

# Antidepressant Agents in Himalayan Beverages

J. P. SHABAAZ BEGUM,[1] K. ASHWINI,[2] LEIRIKA NGANGOM,[1] SHIKHA TOMAR,[1] DIVYA VENUGOPAL,[1] and G. ABHILASH[3]

[1]*Department of Life Sciences, Graphic Era (Deemed to be University), Dehradun, Uttarakhand, India*

[2]*Department of Microbiology, Biotechnology, and Food Technology, Bangalore University, Bangalore, Karnataka, India*

[3]*Bangalore Medical College and Research Institute, Bangalore, Karnataka, India*

**ABSTRACT**

For several decades, depression has emerged as a major mental disorder. According to the report from the World Health Organization (WHO), there are approximately 280 million individuals worldwide suffering from depression. Among them, adults aged over 60 are afflicted at a rate of 5.7%, while adults overall are affected at a rate of 5.0%. In severe cases, depression can cause significant limitations that impede a person's ability to engage in crucial life activities. It is common for mental problems such as depression, anxiety, and sleeplessness to coexist. Depression is a complex interaction of sociological, psychological, and biological factors. Individuals who have experienced tragic events are more likely to develop depression. While there are several medications available worldwide for treating depression, they can have long-lasting side effects. In the past decade, herbal depression medications have gained popularity among

Himalayan Medicinal Plants for the Treatment of Depression: A Source of Rich Antidepressant Agents.
Abdur Rauf, Prabhakar Semwal, & Hafiz Ansar Rasul Suleria (Eds.)

people. The Indian Himalayan region is renowned for its vast variety of ethnomedicinal plants. The local population, with limited access to contemporary medications, uses these plants to cure various illnesses. The plants with health benefits are transformed into palatable staple meals and drinks that are regularly consumed, rather than being used solely as medication. This chapter provides information on the beverages produced and consumed by the tribal communities of the Indian Himalayan Region, which have antidepressant properties.

## 7.1 INTRODUCTION

A mental ailment known as depressive disorder has existed since ancient times. Hippocrates, the Greek physician, described its clinical features and coined the term "melancholia," which is still in use today. Up until the middle of the 20th century, there was no cure for this widespread, severe, and potentially fatal illness. Depression is a progressive and recurrent disorder. 75% of patients experience more than one depressive episode, despite advances in our understanding of its biology and the availability of various pharmaceutical and psychosocial therapies. If the initial episode occurred during youth and there is a family history of depression, the likelihood of recurrence is higher [30]. The longer the illness persists, the poorer the response to treatment and the less likely remission becomes [62]. Long-term antidepressant medication treatment is often necessary for patients with recurring episodes and treatment-resistant chronic depression. This may be due to functional issues in the brain and potential anatomical abnormalities that arise during the disorder and become permanent. Early and effective therapy leading to full remission may improve long-term outcomes and prevent irreversible brain abnormalities. When treating depressive disorders and evaluating the effectiveness of antidepressant medications, it is crucial to consider these factors and both short- and long-term outcomes. Current medication therapies do not address the underlying illness but rather manage symptoms, promoting the body's natural healing process. Anxiety and depression have various root causes and symptoms [10]. Many individuals with mild to moderate conditions use complementary and alternative medicine (CAM) with or without a healthcare professional's recommendation. In developing and emerging economies in East and Southeast Europe, people primarily rely on family members and the media as sources of knowledge [8].

Due to the complex nature of depression, not all patients will exhibit the same anomalies and biomarkers. Future medical procedures will be skilled at locating the precise aberrations unique to each patient independently and focusing on these biomarkers with certain medications. While not all patients have immunological abnormalities, those who do are more likely to benefit from therapies that focus on them. Given the failure of all treatments other than those that now target the monoamine systems, combination therapies may be necessary, depending on the patient's unique profile. These therapies may involve combining a drug that boosts monoamine activity with an anti-inflammatory, a glucocorticoid antagonist, a therapy that targets the glutamate system, or other drugs. As pharmacogenetics develops, we will be better able to individually tailor the antidepressant prescription treatment plan for each patient based on that person's identified abnormalities, genetic features (pharmacokinetic and pharmacodynamic), and the existing medical state. Future research must take into account the fact that depression is often a relapsing-remitting condition, and that prompt treatment is essential for a full recovery.

Some patients may need to take medications for a lifetime; therefore, questions of tolerance and safety are as important as efficacy. However, given our current understanding of the biology of depression, this is a far more challenging and realistically unachievable endeavor. Instead, we should strive to find medications that are curative rather than merely treating symptoms and preventing recurrence.

## 7.2 DEPRESSION

The most common illnesses among the general public and general practitioners are depression and anxiety disorders. Anxiety disorders often co-occur with depression, and patients with depression frequently exhibit symptoms of both disorders. Both disorders can co-occur and meet the criteria for both simultaneously. Due to their association with severe morbidity and mortality, distinguishing between them can be challenging, but it is crucial to recognize and treat both disorders [16]. General practitioners are in a good position to identify and take the lead in treating these conditions to promote better mental health outcomes. According to the National Institutes of Mental Health, depression affects approximately 16 million Americans. Symptoms of depression, a severe mood disorder, include anhedonia (a decreased capacity for enjoyment), sleep disturbances, psychomotor agitation or retardation, fatigue, feelings of guilt or worthlessness, difficulty concentrating, and persistent thoughts of suicide or death. The lack of progress in developing new psychotropic medications by a significant portion of the pharmaceutical industry has contributed to the limited effectiveness of treating major depressive disorder (MDD) [15]. On average, only about one-third of patients with MDD who initiate treatment achieve remission four months later. Depression, a widespread psychiatric problem affecting around 300 million people worldwide, is the fourth most expensive disease, according to WHO [92].

### 7.2.1 DEPRESSION SIGNS AND SYMPTOMS

Some of the signs of depression include worrying, apathy, feeling generally alienated, guilt, despair, loss of interest, losing delight in hobbies, mood fluctuations, and melancholy. Agitation, excessive crying, irritability, restlessness, or social withdrawal are examples of behavioral warning signs. Other warning signs include early awakening, excessive daytime sleepiness, insomnia, or disturbed sleep, excessive hunger, lethargy, appetite loss, inability to concentrate, slowness in activity, or suicidal thoughts, as well as weight gain or loss and repetitive thoughts. Weight gain or loss, persistent pain, heart disease, issues with sexual function, deterioration of long-term medical disorders, difficulty sleeping, gastrointestinal issues, and other consequences are some of the effects of depression. Typically, medication, talk therapy, or a combination of the two are used to treat depression. According to research, these therapies could normalize depressive-related brain alterations. Therapies such as cognitive behavioral therapy and psychotherapy are also available. SSRIs, antidepressants, and anxiolytics are also available as drugs. There is a medical treatment called electroconvulsive therapy.

### 7.2.2 ANTIDEPRESSANTS: THEIR USES

Additionally, these drugs are also used to treat other illnesses in addition to depression. Antidepressants are primarily used to treat the following conditions: depression and MDD, agitation, obsessive-compulsive disorders (OCD), juvenile enuresis or bedwetting, anxiety disorders, bipolar disorder, panic disorder, and post-traumatic stress disorder (PTSD). The usage of a drug "off-label" occurs often. As a result, a doctor may choose to use it even if the FDA has not authorized it, as it may be a beneficial therapy. Sleep issues, discomfort, and migraines are among the off-label uses of antidepressants. According to studies, antidepressants are used for off-label purposes in 29% of cases [63].

However, as we gain more knowledge about depression, more ideas start to surface that help us comprehend its deeper, underlying roots. The majority of contemporary chemical medications are thought to work by increasing the levels of brain-related chemicals associated with neurotransmitters. Serotonin and noradrenaline have been connected to mood and emotion. Uncontrolled use of these chemicals results in detrimental effects. Plants have been used as medicine by the human race since recorded history began. However, traditional medicine use has decreased due to the advancement of modern treatment. Nevertheless, over the last several decades, the use of natural treatments for conditions like depression and anxiety has significantly increased. Herbal remedies are considered a safer alternative to medication since they have a decreased chance of withdrawal symptoms or negative side effects [74]. In Asian countries, the use of traditional remedies is widespread, and there is abundant ancient literature available on the usage of plants or multiple plants widely known as polymedicines for the treatment of depression. Herbal remedies from TCM that have traditionally been used to treat depression include Kai-Xin-San (KXS) [17]. Saiko-ka-ryukotsu-borei-to is another herbal medicine found in Chinese literature used for the treatment of depression and anxiety, for alleviating energy [47]. The Indian Materia Medica, popularly known as Bhavprakasha Nighantu, describes the usage of different plants for treating depression and anxiety. *Bacopa monnieri* is one such plant widely used for improving the cognitive functions of the brain [66]. The majority of research on the topic suggests that Himalayan drinks can reduce the risk of depression, but no studies have utilized the most recent research on the pathophysiology of depression to provide a detailed explanation of the mechanisms underlying these data. To provide in-depth insight into the processes behind the observed antidepressant benefits of both alcoholic and non-alcoholic beverages, this report employs a current conceptual framework of depression neurobiology.

## 7.3 A BRIEF DESCRIPTION OF BEVERAGES

Alcoholic drinks are a significant category of fermented food products that have been prepared and consumed by humans since ancient times in various cultures around the world. A variety of traditional alcoholic beverages are manufactured using local technology and fermentation knowledge in several tribal regions of South Asia. Due to this, homemade alcohol production is a significant sector in rural areas. Based on the variety of raw materials, processing procedures, alcoholic fermentation, distillation, and post-distillation processing, these beverages can come in a wide range of varieties (Figure 7.1) [26]. Native beverage production typically uses natural or spontaneous fermentation, which may include a variety of microorganisms and sugar-rich ingredients such as fruits, grains, cereals, and other foods. Alcoholic beverages are one of the most significant fermented food products, and many of these beverages have been created and consumed by people since ancient times in the majority of the world's cultures [70]. The high-altitude Himalayan area of India is well-known for being the main location for the manufacture of traditional fermented foods and drinks by the local ethnic and tribal populations, who have developed their own cultures for fermentation by making use of abundant natural resources. There are believed to be 250 different types of indigenous fermented foods and beverages that are created by various tribal people in the Himalayan region. The drinking of these ethnic beverages has roots in numerous cultural and religious customs practiced by each tribe group. These traditional foods and drinks are not just a source of household and societal drinking, but also a means of subsistence. The therapeutic value of fermented alcohols, or "Madya," is also mentioned in classical literature such as "Ayurveda" [44].

### *7.3.1 ALCOHOLIC BEVERAGES*

The Himalayan region is well known for producing fermented food and beverages. It is believed that these predigested food and drinks are easily absorbed by the body and help to combat extreme weather conditions. The fermented beverages are highly popular among the tribal communities, and in some cultures, they are produced as offerings to God and later distributed as holy drinks. Additionally, the preparation of these drinks

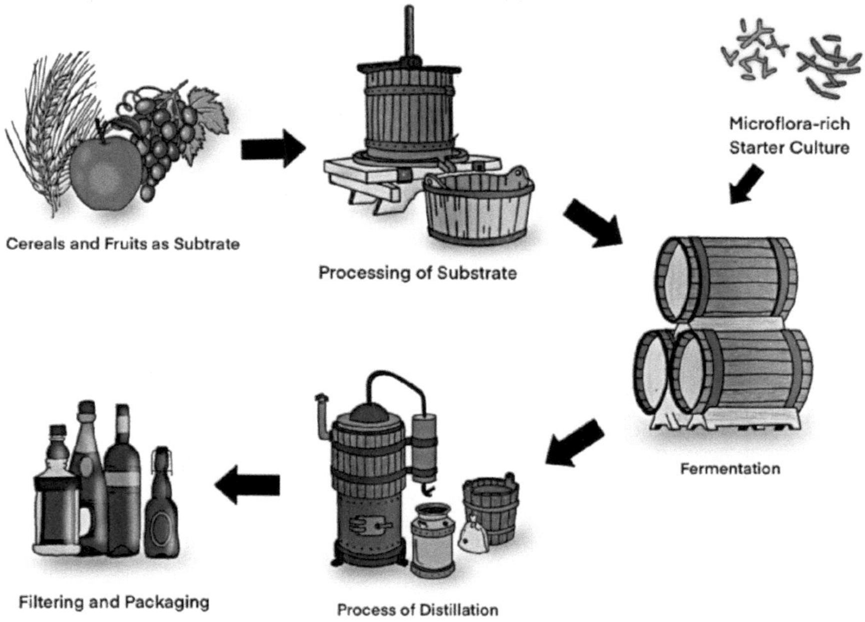


**FIGURE 7.1** Visualization illustrating the process of preparing beverages.

involves the use of various indigenous plants and fruits as raw materials, which, upon fermentation, provide a wide range of bioactive properties to the beverages. The aboriginal communities living in higher altitudes strongly believe that these drinks offer multiple health benefits (as shown in Table 7.1).

### *7.3.2 NON-ALCOHOLIC BEVERAGES*

Around the world, 17% of people suffer from major depression, a prevalent psychiatric disease. According to estimates, depression contributes to over 1 million suicides annually, placing a significant burden on society. Even though antidepressants have been medically available for a long time, the majority of them are not completely effective (only approximately 33% of the total depressed patients respond favorably to the first antidepressant treatment) and are associated with several negative side effects. Traditional methods of preparing herbs for antidepressants have been the focus of recent studies. As additional interventions to promote health and treat

**TABLE 7.1** List of Beverages and Their Biological Activities

| SL. No. | Name of Beverage | Region | Plant | Medicinal Properties Due to Biological Activity | References |
|---|---|---|---|---|---|
| 1. | Chhang | Nubra valley, Leh-Ladhak | *Artemisia* | Antioxidant, antidepressant properties. | [6] |
| 2. | Jaan | Kumaon region, Uttarakhand | *Cinnamomum zeylanicum* | Antidepressant, antibacterial, antifungal, antioxidant, anti-parasitic activity, lowering of blood glucose, blood pressure and serum cholesterol, and Alzheimer's disease. | [5] |
| | | | *Ammomum subulatum* | Antianxiety and sedative activity. | [20] |
| | | | *Piper longum* | Antianxiety and depression, chronic bronchitis, asthma, constipation, gonorrhea, paralysis of the tongue, diarrhea, cholera, chronic malaria, viral hepatitis, respiratory infections, stomachache, bronchitis, diseases of the spleen, cough, and tumors. | [31] |
| | | | *Ficus religiosa* | Antidepressant, antiulcer, antibacterial, antidiabetic, gonorrhea, and skin diseases. | [13] |
| 3. | Ghenti | Chakrata region, Uttarakhand | *Achyranthes aspera* L. | Antidepressant and sleep inducing. | [28] |
| | | | *Adhatoda zeylanica zeylinica Medik.* | Antimicrobial, anti-inflammatory, and anti-depressant activity. | [39] |
| | | | *Aerva sanguinolenta* (L.) *Bl.* | Anxiolytic, antihyperglycemic, hepatoprotective, antibacterial, antifungal, antioxidant, anti-inflammatory, diuretic, and anti-leishmania. | [68] |
| | | | *Artemisia roxburghiana Wall. ex Bess.* | Antidepressant, antimicrobial, and antioxidant. | [52] |
| | | | *Berberis lycium Royle* | Antidepressant and antioxidant. | [32] |
| | | | *Boerhaavia diffusa* L. | Antistress, anticonvulsant activities, jaundice, asthma, rheumatism, nephrological disorders, ascites, anemia, and gynecological disorders. | [23] |

**TABLE 7.1** *(Continued)*

| SL. No. | Name of Beverage | Region | Plant | Medicinal Properties Due to Biological Activity | References |
|---|---|---|---|---|---|
| | | | *Cannabis sativa* L. | Mood-elevating, sedative, rheumatism, epilepsy, asthma, skin burns and pain. | [25] |
| | | | *Carissa opaca Stapf ex Haines* | Asthma and pulmonary damage, sleep inducing. | [67] |
| | | | *Cassia tora* L. *Cassia tora* | Anxiolytic, antidiabetic, and antioxidant activities. | [43] |
| | | | *Ficus benghalensis L.* | Memory improving, anxiety-reducing, and depression-resolving. | [53] |
| | | | *Melia azedarach* L. | Anxiolytic and anti-depressant. | [33] |
| | | | *Punica granatum* L. | Depressive disorders, menopause. | [82] |
| | | | *Rhus parviflora Roxb.* | Induces sleep. | [73] |
| | | | *Syzygium cumini* L. | Anthelmintic, sore throat, bronchitis, asthma, thirst, biliousness, dysentery, and ulcers, blood purifier. | [7] |
| | | | *Vitex negundo* L. | Anti-inflammatory, expectorant, tranquilizer, antispasmodic, anti-convalescent, rejuvenative, antiarthritic, anthelminthic, antifungal, and antipyretic. | [9] |
| | | | *Woodfordia fruticosa* (L.) *Kurz.* | Reduces hyperglycemia, cellular toxicity, and depressive disorder. | [79] |
| | | | *Zanthoxylum armatum DC.* | Anxiolytic, sleep inducing, spasmolytic effects, gastrointestinal, respiratory, cardiovascular disorders, asthma, bronchitis, indigestion, varicose veins, and diarrhea. | [88] |
| 4. | Lugri | Kullu region, Himachal Pradesh | *Viburnum grandifolia* | Digestive problems, purgative, abdominal pain, diuretic, antimalarial, and muscle relaxant. | [72] |
| | | | *Arisaema tortuosum* | Resolves phlegm problem, dampness, and to treat asthma, cold, cough, bronchitis, and laryngitis. | [1] |

**TABLE 7.1** *(Continued)*

| SL. No. | Name of Beverage | Region | Plant | Medicinal Properties Due to Biological Activity | References |
|---|---|---|---|---|---|
| | | | *Viola canescens* | Sedative, cough, cold, flu, fever, and malaria and anticancer. | [54] |
| | | | *Verbascum thapsus* | Relaxant, antispasmodic, and anti-tubercular agent. | [2] |
| | | | *Bupleurum lanceolatum/ Valeriana jatamansi* | Sedative, analeptic, antispasmodic, carminative, stimulant, stomachic, and nervine. | [57] |
| | | | *Swertia chirayata* | General stress, liver disorders, malaria, and diabetes. | [42] |
| | | | *Centella asiatica* | Antidepressant and antimicrobial. | [86] |
| | | | *Solanum xanthocarpum* | Antioxidant, antimalarial, and antidepressant activity. | [38] |
| | | | *Clitoria ternatea* | Memory enhancer, antidepressant, antistress, anxiolytic, tranquilizing, anticonvulsant, and sedative agent. | [58] |
| | | | *Aegel marmelos* | Anxiolytic and antidepressant. | [41] |
| | | | *Trachyspermum ammi* | Antidepressant and muscle relaxant. | [65] |
| | | | *Salvia* sp. | Anxiolytic, anti-depressant, anticancer, anti-inflammatory, antinociceptive, antioxidant, antibacterial, antifungal, antimutagenic, antidementia, hypoglycemic, and hypolipidemic effects. | [29] |
| | | | *Fragaria* sp. | Antidepressant, sleep inducing, laxative, and diuretic. | [61] |

diseases, natural compounds, particularly plant polyphenols, have drawn increasing amounts of attention. Many different flavored teas have traditionally been recognized as sedatives such as kava, valerian, chamomile, passionflower, and St. John's wort [12].

#### *7.3.2.1 GREEN TEA*

One of the most popular drinks in the world is tea, a byproduct of the plant Camellia sinensis (L.) O. Kuntze. Regardless of the degree of fermentation, the same plant is utilized to produce green tea, black tea, and oolong tea [56, 76]. Among them, the health benefits of green tea have garnered the most attention, including its anti-cancer, anti-obesity, anti-diabetes, anti-inflammatory, and anti-neurodegenerative activities [89]. Green, oolong, and black tea are all derived from the leaves of Camellia sinensis using full, semi-, and non-fermentation processes, respectively. Green tea was first discovered in China around 3000 BC, and its therapeutic properties have been recognized for a long time [77, 89]. Green tea contains various chemical compounds such as theanine, caffeine, and green tea catechins (GTCs), which may influence brain function. Chen et al. extensively reviewed the neuroprotective properties and mechanisms of action of different tea additives, such as theaflavins, caffeine, theanine, and catechins. They suggested that these bioactive tea components could potentially aid in the treatment of neuronal degeneration [14].

The unfavorable correlations between green tea consumption and the risk of depression have been explained by several different processes. Green tea is abundant in catechins, a type of polyphenol [48]. In a mouse model of depression, the oral treatment of green tea polyphenol for seven days greatly decreased immobility, indicating that this polyphenol has antidepressant-like properties [46]. Green tea polyphenols also decreased corticosterone and adrenocorticotropic hormone levels, which inhibited the hypothalamic-pituitary-adrenal axis and decreased unfavorable reactions to stress [91]. One of the main causes of depressive illnesses and one of the most well-known is stress.

Recent studies have suggested that the amino acid theanine, which is present in green tea [48], can alleviate psychological stress and have neuroprotective effects by regulating neurotransmitters. Theanine has been shown to have an antidepressant effect in animal investigations

[81]. The concentrations of numerous brain neurotransmitters, such as dopamine and serotonin, may be regulated by theanine [60], and depression may result from their malfunction [22]. The effects of tea components on psychological stress and anxiety have been studied in several animal experiments [45]. Anxiolytic and stress-reducing effects may aid in the alleviation of depressive symptoms. Specific animal models of depression are mostly chosen based on their capacity to mimic the effects of well-known antidepressant drugs in humans. The most often used models make use of behavioral paradigms like learned helplessness, social defeat, and long-term moderate stress. Behavioral testing for antidepressants in rodents is usually done using the forced-swim test and the tail-suspension test. In mouse models of behavioral depression, for instance, tea polyphenols [91] and theanine [81] have been demonstrated to be beneficial. Theanine's antagonistic effects on ptosis and hypothermia brought on by reserpine suggest an effect mediated by monoamine neurotransmitters [90]. In another study, an antidepressant effect was also discovered as evidenced by a reduction in immobility during the forced swim test (FST) [83]. In research on humans, the antidepressant properties of tea components are at most modest. Particularly, mood improvement in otherwise healthy persons shouldn't cause one to overestimate potential therapeutic improvements in those with major depressive illness. If positive effects on mood can be demonstrated, tea drinking could develop into a cost-effective method of promoting mental health, for instance among older persons. Randomized controlled intervention studies are required to establish a causal link between depression and the bioactive chemicals found in tea.

### *7.3.2.2 CHAMOMILE TEA*

Ancient Greece and Rome were among the first civilizations to embrace chamomile as a plant cure. It has a long history of usage in traditional medicine as an anti-inflammatory, antioxidant, mildly astringent, and healing agent. It has been used often as an aqueous extract to treat various gastrointestinal disorders, as a moderate sedative to ease anxiety and calm the nerves, and to treat hysteria, nightmares, insomnia, and other sleep issues [75]. Despite this historical use, there aren't many studies that examine how chamomile affects the central nervous system

(CNS). In controlled clinical trials, it was found that chamomile dry extracts (standardized to contain 1.2% apigenin) may have both anti-depressant and anxiolytic effects on people with mild to moderate generalized anxiety disorder (GAD) [3, 37]. Based on scant evidence, chamomile and certain of its flavonoid constituents may have anxiolytic and antidepressant properties. Anxiety disorders, sadness, and poor sleep quality can all be treated with chamomile tea and the flavonoids it contains, particularly apigenin, which binds to benzodiazepine receptors in the brain. Patients who can't take SSRIs or SNRIs (selective serotonin-reuptake inhibitors or serotonin-norepinephrine (NE) reuptake inhibitors), prefer natural remedies, or refuse conventional pharmaceuticals due to stigma and other socio-cultural factors may find the side effect profile of chamomile to be particularly appealing. Flavonoids found in chamomile have benzodiazepine-like effects as well as phosphodiesterase inhibitory properties, which raise cAMP levels [24]. Chamomile tea works by inhibiting phosphodiesterase, which raises the level of cAMP. A recent study compares *Matricaria recutita* (L) extract to placebo for eight weeks in Indonesian individuals with mild to moderate GAD (DSM-IV). Colds, sore throats, abscesses, gum inflammation (gingivitis), anxiety, sleeplessness, acne, eczema, minor burns, inflammatory bowel disease (ulcerative colitis), ulcers, and children's illnesses including chickenpox, diaper rash, and colic have all traditionally been treated with chamomile [12].

Researchers examined the impact of chamomile tea on cancer patients undergoing chemotherapy, specifically for anxiety and depression. The study revealed that consuming chamomile tea once a day for two weeks had significant positive effects on depression, but did not affect anxiety levels in cancer patients. Due to its minimal side effects, safety, ease of use, and affordability, chamomile tea can be considered as an alternative therapy for depression in cancer patients undergoing chemotherapy. It is recommended that cancer patients undergoing chemotherapy incorporate chamomile tea into their routine as an effective strategy to alleviate depressive symptoms [27]. Although the anxiolytic activity of chamomile is not well understood, research suggests that one or more of its flavonoid components may act as an antidepressant by modulating neurotransmission of central noradrenaline, dopamine, serotonin, and aminobutyric acid [27].

#### 7.3.2.3 WARM MILK

The Himalayan highlands are home to a variety of cow breeds that may provide nutrient-rich milk. The majority of Indians are familiar with the ancient practice of consuming hot milk before bed to aid in falling asleep. Tryptophan, an amino acid found in milk, is converted into a monoamine neurotransmitter, a clever neurochemical with a soothing effect that helps elevate mood and promote relaxation. Additionally, the pleasant temperature of the milk has a cooling and calming effect on the body. When you are exhausted from a long day of work and seeking restful sleep, this can be highly beneficial [80].

Tryptophan, one of the essential amino acids in milk, is crucial for the production of serotonin. By raising awareness of nutritional requirements for the elderly and minimizing cognitive decline, the quality of life for older individuals can be significantly improved [24].

#### 7.3.2.4 TART CHERRY JUICE

Cherry is a member of the Rosaceae family, more specifically the Prunus genus. The two most popular varieties of Prunus are *Prunus cerasus* and *Prunus avium*, commonly known as sour cherry and sweet cherry, respectively. Both varieties are considered nutrient-dense foods due to their low-calorie counts and high concentrations of vital nutrients and bioactive dietary ingredients [55]. Numerous studies have shown that a diet high in fruit can reduce the risk of oxidative stress, diabetes, cancer, cardiovascular disease, and neurological illnesses [21, 87].

Research indicates that tart cherry juice affects two primary types of enzymes: those involved in glucose metabolism and those involved in the central nervous system (CNS). For the first time, tart cherry juice has been found to successfully inhibit TYR (Tyrosinase) and MAO-A (monoamine oxidase A). Certain polyphenols, such as anthocyanins, are believed to be responsible for the suppression of MAO-A, which can have antidepressant and anxiolytic effects by preventing the deamination of catecholamines and serotonin [59]. Tart cherries have higher anthocyanin concentrations compared to sweet cherries and other fruits. There is a positive correlation between cherry anthocyanin concentrations and neuronal cell protection against oxidative stress [84,

85]. The presence of anti-inflammatory cytokines in tart cherries, which are closely linked to sleep regulation, suggests a potential approach to enhancing sleep [64]. Additionally, certain cherries have relatively high melatonin concentrations, which may act as a source of exogenous melatonin, a chemical known for its sleep-regulating properties. However, there is a lack of clinical studies investigating the effects of cherries on sleep.

#### *7.3.2.5 KAVA TEA*

The *Piper methysticum* plant, which is indigenous to the tropical islands of Polynesia, is commonly used for its relaxing and uplifting properties. The active components (dihydrokawain, methysticin, dihydromethysticin, and kawain) of *P. methysticum's* root extract are known as Kava lactones. Historically, its usage has been limited to Polynesian culture; however, the potential side effects have sparked interest in the medical community for its introduction into the Western market as a health supplement [51]. It spread to the eastern part of Polynesia following interaction with the Austronesian Lapita civilization. Kava is exclusive to Oceania and not found in other Austronesian populations. While it made its way to Hawaii, it is absent in New Zealand due to its inability to grow there. The use of Kava is also believed to be the reason why Austronesians in Oceania stopped chewing betel, a common practice elsewhere. Research suggests that the reconstructed Proto-Polynesian term for the plant, kava, comes from the Proto-Oceanic word kawaR, which means a bitter or powerful root (used as fish poison) [50]. It may have undergone metathesis with the newly discovered wakaR (in Proto-Oceanic and Proto-Malayo-Polynesian). Originally, it referred to Zingiber zerumbet, which was used in Austronesian rituals to prepare a similar bitter beverage with mild hallucinogenic effects. Kava lactones, the active compounds in kava, have similar effects on brain chemistry as antidepressants and anti-anxiety medications. Kavain, a kavalactone, is particularly effective in reducing anxiety. By providing temporary relief from worry, tension, and sleeplessness, kava helps to relax the mind while maintaining clarity. Essentially, kava acts as a depressant, slowing down the transmission of signals between the brain and the body [69].

#### *7.3.2.6 LEMON BALM TEA*

A member of the Lamiaceae family, lemon balm (Melissa officinalis L.) is a perennial herb. This plant is utilized in traditional medicine of several cultures to treat a wide range of ailments. Recent pharmacological research has supported empirical data regarding its use in traditional medicine [71]. Lemon balm has been used as a memory enhancer, heart tonic, antidepressant, sleeping aid, and antidote for various malignancies, mental and CNS disorders, circulatory and respiratory issues, and cardiovascular illnesses [19]. In Asian traditional medicine, lemon balm has been successfully used to treat many mental illnesses including depression, anxiety, sleeplessness, anxiety-induced palpitations, and stress [71]. Research conducted on the CNS system has shown that lemon balm has neuroprotective, anti-insomnia, anti-depressant, and anxiolytic properties. These studies have also examined the effects of lemon balm on mood, cognitive function, and memory [71, 49]. The Food and Drug Administration (FDA) has declared lemon balm to be GRAS (generally recognized as safe) [11]. Previous animal experiments have shown that fluoxetine was more effective than lemon balm in reducing depressed behaviors in non-stressed rats [78]. However, in a recent scientific experiment, no noticeable difference was observed between the antidepressant effects of fluoxetine and lemon balm [4].

#### *7.3.2.7 ASHWAGANDHA TEA*

Ashwagandha is a member of the Solanaceae family. It is also known as Indian ginseng, poison gooseberry, and winter cherry. Ashwagandha is cultivated in the northwestern and central regions of India, primarily in the states of Madhya Pradesh, Gujarat, Haryana, Maharashtra, Punjab, Rajasthan, and Uttar Pradesh. It is also found in Yemen, China, and Nepal [40, 18]. Extracts from the roots, berries, and leaves of the ashwagandha plant have been used to treat stress, anxiety, and arthritis. Studies on ashwagandha leaf powder extract have shown that it possesses anti-inflammatory, anti-apoptotic, and anti-anxiety properties, which may help mitigate the negative effects of obesity and related diseases [35]. Ashwagandha has a long history of use in Ayurvedic medicine. The active ingredients in the root can aid in sleep when extracted and used in high doses. Ashwagandha leaves contain trimethylene glycol, a substance that promotes the body's sleep cycle and tissue and bone renewal [36]. While ashwagandha tea is generally safe, caution should be exercised by individuals with autoimmune diseases, pregnant or nursing women, and those taking medications for high blood pressure, diabetes, or thyroid disease. It is always advisable to consult a doctor to determine if it is suitable for your health. In addition to potentially calming the mind, reducing edema, lowering blood pressure, and modulating the immune system, ashwagandha contains compounds that may provide assistance. Ashwagandha is commonly used as an adaptogen to treat various stress-related conditions. Adaptogens are believed to strengthen the body's defenses against physical and emotional stress. One of the most well-known uses of ashwagandha is for stress relief. It is classified as an adaptogen, a substance that helps alleviate the effects of stress on the body. Ashwagandha appears to assist in managing stress mediators such as cortisol, heat shock proteins (Hsp70), and stress-activated c-Jun N-terminal protein kinase (JNK-1) [34].

## 7.4 SUMMARY

An important mental condition that has impacted individuals of all ages worldwide is depression. Although there are many different medications available to treat depression, they all have negative side effects. Moreover, antidepressant addiction may develop with continued use. Many individuals consider using herbal medications for the treatment of depression to avoid these unwanted effects. The tribal people of the Indian Himalayan area are also well-known for producing several drinks that they use to alleviate stress, sleeplessness, and anxiety by utilizing ethnomedical herbs. To enhance mindfulness and improve overall mental health, people consume ashwagandha tea, chamomile tea, green tea, and other decoctions. Additionally, it is widely known that indigenous people consume fermented drinks made from a blend of herbs. Each community is known to create its own starter culture to aid in the fermentation process. Plants and flour are primarily used in the preparation of the starter culture. Although there is ample evidence supporting the antidepressant properties of the herbs used in the creation of starter cultures, further investigation is still required to demonstrate the efficacy of the drink as an antidepressant.

## KEYWORDS


- antidepressant
- anxiety
- herbal medications
- herbs
- Himalayan beverages
- Indian Himalayan region
- medicinal plants
- mental disorders


## REFERENCES


1. Ali, H., & Yaqoob, U., (2021). Traditional uses, phytochemistry, pharmacology and toxicity of *Arisaema* (*Areaceae*): A review. *Bulletin of the National Research Centre, 45*(1), 1–19.
2. Ali, N., Ali Shah, S.- W., Shah, I., Ahmed, G., Ghias, M., Khan, I., & Ali, W., (2012). Anthelmintic and relaxant activities of *Verbascum thapsus* mullein. *BMC Complementary and Alternative Medicine, 12*(1), 1–7.
3. Amsterdam, J. D., Li, Y., Soeller, I., Rockwell, K., Mao, J. J., & Shults, J., (2009). A randomized, double-blind, placebo-controlled trial of oral *Matricaria recutita* (chamomile) extract therapy of generalized anxiety disorder. *Journal of Clinical Psychopharmacology, 29*(4), 378.
4. Araj-Khodaei, M., Noorbala, A. A., Yarani, R., Emadi, F., Emaratkar, E., Faghihzadeh, S., & Naseri, M., (2020). A double-blind, randomized pilot study for comparison of *Melissa officinalis* L. and *Lavandula angustifolia* Mill. with fluoxetine for the treatment of depression. *BMC Complementary Medicine and Therapies, 20*(1), 1–9.
5. Aryanezhad, M., Abdi, M., Amini, S., Hassanzadeh, K., Valadbeigi, E., Rahimi, K., & Moloudi, M. R., (2021). *Cinnamomum zeylanicum* extract has antidepressant-like effects by increasing brain-derived neurotrophic factor (BDNF) and its receptor in prefrontal cortex of rats. *Avicenna Journal of Phytomedicine, 11*(3), 302.
6. Asgharzadeh, N., Lorigooini, Z., Amini-Khoei, H., Ghaderian, A., Mardani, M., Moradi, M. T., & Shahrani, M., (2020). Anti-anxiety effects of *Artemisia persica* in male rats. *Research Journal of Pharmacognosy, 7*(4), 65–73.
7. Ayyanar, M., & Subash-Babu, P., (2012). *Syzygium cumini* (L.) skeels: A review of its phytochemical constituents and traditional uses. *Asian Pacific Journal of Tropical Biomedicine, 2*(3), 240–246.

8. Ballenger, J. C., Davidson, J. R., Lecrubier, Y., Nutt, D. J., Borkovec, T. D., Rickels, K., & Wittchen, H. U., (2001). Consensus statement on generalized anxiety disorder from the international consensus group on depression and anxiety. *Journal of Clinical Psychiatry*, *62*, 53–58.
9. Bano, U., Jabeen, A., Ahmed, A., & Siddiqui, M. A., (2015). Therapeutic uses of *Vitex negundo*. *World Journal of Pharmaceutical Research*, *4*(12), 589–606.
10. Barlow, D. H., & Campbell, L. A., (2000). Mixed anxiety-depression and its implications for models of mood and anxiety disorders. *Comprehensive Psychiatry*, *41*(2), 55–60.
11. Basch, E., Bent, S., Foppa, I., Haskmi, S., Kroll, D., Mele, M., & Yong, S., (2006). Marigold (*Calendula officinalis* L.) an evidence-based systematic review by the natural standard research collaboration. *Journal of Herbal Pharmacotherapy*, *6*(3, 4), 135–159.
12. Beaubrun, G., & Gray, G. E., (2000). A review of herbal medicines for psychiatric disorders. *Psychiatric Services*, *51*(9), 1130–1134.
13. Chandrasekar, S. B., Bhanumathy, M., Pawar, A. T., & Somasundaram, T., (2010). Phytopharmacology of *Ficus religiosa*. *Pharmacognosy Reviews*, *4*(8), 195.
14. Chen, S. Q., Wang, Z. S., Ma, Y. X., Zhang, W., Lu, J. L., Liang, Y. R., & Zheng, X. Q., (2018). Neuroprotective effects and mechanisms of tea bioactive components in neurodegenerative diseases. *Molecules*, *23*(3), 512.
15. Cowen, P. J., (2017). Backing into the future: Pharmacological approaches to the management of resistant depression. *Psychological Medicine*, *47*(15), 2569–2577.
16. Cuijpers, P., & Smit, F., (2002). Excess mortality in depression: A meta-analysis of community studies. *Journal of Affective Disorders*, *72*(3), 227–236.
17. Dang, H., Sun, L., Liu, X., Peng, B., Wang, Q., Jia, W., & Xiao, P., (2009). Preventive action of Kai Xin San aqueous extract on depressive-like symptoms and cognition deficit induced by chronic mild stress. *Experimental Biology and Medicine*, *234*(7), 785–793.
18. Dar, N. J., Hamid, A., & Ahmad, M., (2015). Pharmacologic overview of *Withania somnifera*, the Indian ginseng. *Cellular and Molecular Life Sciences*, *72*(23), 4445–4460.
19. Dastmalchi, K., Dorman, H. D., Oinonen, P. P., Darwis, Y., Laakso, I., & Hiltunen, R., (2008). Chemical composition and *in vitro* antioxidative activity of a lemon balm (*Melissa officinalis* L.) extract. *LWT-Food Science and Technology*, *41*(3), 391–400.
20. Deepanchakkaravarthi, S., (2017). *Evaluation of Anti-anxiety and Sedative Effects of Various Extracts of Amomum Subulatum Seeds in Swiss Albino mice*. Doctoral dissertation, Madras Medical College, Chennai.
21. Del Bo, C., Martini, D., Porrini, M., Klimis-Zacas, D., & Riso, P., (2015). Berries and oxidative stress markers: An overview of human intervention studies. *Food & Function*, *6*(9), 2890–2917.
22. Delgado, P. L., (2000). Depression: The case for a monoamine deficiency. *Journal of Clinical Psychiatry*, *61*(6), 7–11.
23. Dhingra, D., & Valecha, R., (2014). Evidence for involvement of the monoaminergic system in antidepressant-like activity of an ethanol extract of *Boerhaavia diffusa* and its isolated constituent, punarnavine, in mice. *Pharmaceutical Biology*, *52*(6), 767–774.

24. Effendy, E., Sembiring, I. M. B., & Utami, N., (2019). The effectiveness of warm milk and chamomile tea on anxiety symptoms in the elderly. *Proceedings of the International Conference on Natural Resources and Technology (ICONART, 2019)*, 322–326.
25. El-Alfy, A. T., Ivey, K., Robinson, K., Ahmed, S., Radwan, M., Slade, D., & Ross, S., (2010). Antidepressant-like effect of Δ9-tetrahydrocannabinol and other cannabinoids isolated from *Cannabis sativa* L. *Pharmacology Biochemistry and Behavior*, *95*(4), 434–442.
26. Franz, C. M., Huch, M., Abriouel, H., Holzapfel, W., & Gálvez, A., (2011). Enterococci as probiotics and their implications in food safety. *International Journal of Food Microbiology*, *151*(2), 125–140.
27. Ghamchini, V. M., Salami, M., Mohammadi, G. R., Moradi, Z., Kavosi, A., Movahedi, A., & Aryaeefar, M. R., (2019). The effect of chamomile tea on anxiety and depression in cancer patients treated with chemotherapy. *Journal of Young Pharmacists*, *11*(3), 309.
28. Goel, R. K., Gawande, D. Y., Lagunin, A. A., & Poroikov, V. V., (2018). Pharmacological repositioning of *Achyranthes aspera* as an antidepressant using pharmacoinformatic tools PASS and PharmaExpert: A case study with wet lab validation. *SAR and QSAR in Environmental Research*, *29*(1), 69–81.
29. Guenné, S., Kinda, P. T., Lefter, R., Timofte, D., Hilou, A., Cantemir, A., & Kiendrebéogo, M., (2021). Anxiolytic, anti-depressant and metabolic effect of *Salvia* spp. essential oil on a rat model of schizophrenia. *Bulletin of Integrative Psychiatry*, *27*(2), 59–66.
30. Hollon, S. D., Shelton, R. C., Wisniewski, S., Warden, D., Biggs, M. M., Friedman, E. S., & Rush, A. J., (2006). Presenting characteristics of depressed outpatients as a function of recurrence: Preliminary findings from the STAR* D clinical trial. *Journal of Psychiatric Research*, *40*(1), 59–69.
31. Hritcu, L., Noumedem, J. A., Cioanca, O., Hancianu, M., Postu, P., & Mihasan, M., (2015). Anxiolytic and antidepressant profile of the methanolic extract of *Piper nigrum* fruits in beta-amyloid (1–42) rat model of Alzheimer's disease. *Behavioral and Brain Functions*, *11*(1), 1–13.
32. Iqbal, J., Khan, A., Muhammad, S., Jan, S. U., & Gul, R., (2020). Analgesic and neuropharmacological activities of *Berberis lycium* royle collected from district Sherani Balochistan, Pakistan. *Pakistan Journal of Pharmaceutical Sciences*, *33*(4).
33. Ishaq, H., (2016). Anxiolytic and antidepressant activity of different methanolic extracts of *Melia azedarach* Linn. *Pakistan Journal of Pharmaceutical Sciences*, *29*(5).
34. Kaur, A., Singh, B., Ohri, P., Wang, J., Wadhwa, R., Kaul, S. C., & Kaur, A., (2018). Organic cultivation of ashwagandha with improved biomass and high content of active withanolides: Use of vermicompost. *PLoS One*, *13*(4), e0194314.
35. Kaur, T., & Kaur, G., (2017). *Withania somnifera* as a potential candidate to ameliorate high fat diet-induced anxiety and neuroinflammation. *Journal of Neuroinflammation*, *14*(1), 1–18.
36. Kaushik, M. K., Kaul, S. C., Wadhwa, R., Yanagisawa, M., & Urade, Y., (2017). Triethylene glycol, an active component of Ashwagandha (*Withania somnifera*) leaves, is responsible for sleep induction. *PLoS One*, *12*(2), e0172508.

37. Keefe, J. R., Mao, J. J., Soeller, I., Li, Q. S., & Amsterdam, J. D., (2016). Short-term open-label chamomile (*Matricaria chamomilla* L.) therapy of moderate to severe generalized anxiety disorder. *Phytomedicine*, *23*(14), 1699–1705.
38. Khan, A. W., & Khan, A. U., (2019). Antidepressant effect of *Solanum surattense* Burm. F. *The Journal of Animal and Plant Sciences*, *29*(4), 1188–1192.
39. Khursheed, A., Devender, P., & Ansari, S. H., (2010). Phytochemical and pharmacological investigations on adhatoda zeylanica (medic.): A review. *Pharmacognosy Journal*, *2*(12), 513–519.
40. Kulkarni, S. K., & Dhir, A., (2008). *Withania somnifera*: An Indian ginseng. *Progress in Neuro-psychopharmacology and Biological Psychiatry*, *32*(5), 1093–1105.
41. Kumar, S., Mahaseth, R. K., Tiwari, M., Sehgal, R., Rajora, P., & Mathur, R., (2015). Interaction of aqueous leaf extract of *Aegle marmelos* (L.) Corr. with cholinergic, serotonergic, and adrenergic receptors: An ex vivo study. *Indian Journal of Pharmacology*, *47*(1), 109.
42. Kumar, V., & Van, S. J., (2016). A review of *Swertia chirayita* (*Gentianaceae*) as a traditional medicinal plant. *Frontiers in Pharmacology*, *6*, 308.
43. Kumar, V., Singh, R., Mahdi, F., Mahdi, A. A., & Singh, R. K., (2017). Experimental validation of antidiabetic and antioxidant potential of *Cassia tora* (L.): An indigenous medicinal plant. *Indian Journal of Clinical Biochemistry*, *32*(3), 323–328.
44. Kumari, A., Pandey, A. N. I. T. A., Ann, A., Raj, A., Gupta, A. N. U. P. A. M. A., Chauhan, A. R. J. U. N., & Jaiswal, V., (2016). Indigenous alcoholic beverages of South Asia. *Indigenous Alcoholic Beverages of South Asia* (pp. 501–566). CRC Press, New York.
45. Lardner, A. L., (2014). Neurobiological effects of the green tea constituent theanine and its potential role in the treatment of psychiatric and neurodegenerative disorders. *Nutritional Neuroscience*, *17*(4), 145–155.
46. Levites, Y., Weinreb, O., Maor, G., Youdim, M. B., & Mandel, S., (2001). Green tea polyphenol (–)-epigallocatechin-3-gallate prevents N-methyl-4-phenyl-1, 2, 3, 6-tetrahydropyridine-induced dopaminergic neurodegeneration. *Journal of Neurochemistry*, *78*(5), 1073–1082.
47. Li, C., Huang, J., Cheng, Y. C., & Zhang, Y. W., (2020). Traditional Chinese medicine in depression treatment: From molecules to systems. *Frontiers in Pharmacology*, *11*, 586.
48. Liao, S., Kao, Y. H., & Hiipakka, R. A., (2001). Green tea: Biochemical and biological basis for health benefits. *Vitamins and Hormones, 1*(62), 1–94.
49. López, V., Martín, S., Gómez-Serranillos, M. P., Carretero, M. E., Jäger, A. K., & Calvo, M. I., (2009). Neuroprotective and neurological properties of *Melissa officinalis*. *Neurochemical Research*, *34*(11), 1955–1961.
50. Lynch, J., (2002). Potent roots and the origin of kava. *Oceanic Linguistics*, 493–513.
51. Eriksson, M., Lindström, B., & Lindstro, B., (2005). Interactions between herbal medicines and prescribed drugs. *Neurol. (English Ed.)*, *8198*(3), 2163–2175.
52. Mahmoudi, M., Ebrahimzadeh, M. A., Ansaroudi, F., Nabavi, S. F., & Nabavi, S. M., (2009). Antidepressant and antioxidant activities of *Artemisia absinthium* L. at flowering stage. *African Journal of Biotechnology*, *8*(24).
53. Malik, H., Javaid, S., Fawad, R. M., Samad, N., Rizwan, A. S., Alqahtani, F., & Imran, I., (2020). Amelioration of scopolamine-induced amnesic, anxiolytic and

antidepressant effects of *Ficus benghalensis* in behavioral experimental models. *Medicina, 56*(3), 144.

54. Masood, M., Arshad, M., Asif, S., & Chaudhari, S. K., (2014). *Viola canescens*: Herbal wealth to be conserved. *Journal of Botany*.
55. McCune, L. M., Kubota, C., Stendell-Hollis, N. R., & Thomson, C. A., (2010). Cherries and health: A review. *Critical Reviews in Food Science and Nutrition, 51*(1), 1–12.
56. Miyoshi, N., Pervin, M., Suzuki, T., Unno, K., Isemura, M., & Nakamura, Y., (2015). Green tea catechins for well-being and therapy: Prospects and opportunities. *Botanics: Targets and Therapy, 5*, 85–96.
57. Muhammad, A., Shaheen, B., Abdul, L. K., Abdul, M. T., & In-Jung, L., (2012). Medicinal plants and their uses in selected temperate zones of Pakistani Hindukush-Himalaya. *Journal of Medicinal Plants Research, 6*(24), 4113–4127.
58. Mukherjee, P. K., Kumar, V., Kumar, N. S., & Heinrich, M., (2008). The ayurvedic medicine *Clitoria ternatea*—From traditional use to scientific assessment. *Journal of Ethnopharmacology, 120*(3), 291–301.
59. Nabavi, S. M., Daglia, M., Braidy, N., & Nabavi, S. F., (2017). Natural products, micronutrients, and nutraceuticals for the treatment of depression: A short review. *Nutritional Neuroscience, 20*(3), 180–194.
60. Nathan, P. J., Lu, K., Gray, M., & Oliver, C., (2006). The neuropharmacology of L-theanine (N-ethyl-L-glutamine) a possible neuroprotective and cognitive enhancing agent. *Journal of Herbal Pharmacotherapy, 6*(2), 21–30.
61. Naz, S., Farooq, U., Khan, A., Khan, H., Karim, N., Sarwar, R., & Rauf, A., (2017). Antidepressent effect of two new benzyl derivatives from wild strawberry *Fragaria vesca* var. nubicola Lindl. ex-Hook. f. *Frontiers in Pharmacology, 8*, 469.
62. Okuda, A., Suzuki, T., Kishi, T., Yamanouchi, Y., Umeda, K., Haitoh, H., & Iwata, N., (2010). Duration of untreated illness and antidepressant fluvoxamine response in major depressive disorder. *Psychiatry and Clinical Neurosciences, 64*(3), 268–273.
63. Olden, K. W., (2005). The use of antidepressants in functional gastrointestinal disorders: New uses for old drugs. *CNS Spectrums, 10*(11), 891–896.
64. Opp, M. R., (2004). Cytokines and sleep: The first hundred years. *Brain, Behavior, and Immunity*.
65. Rahman, M. R., Ali, M., Sharif, M., Sajon, S. R., Mannan, M. A., & Shahed-Al-Mahmud, M., (2018). Antidepressant-like activity of methanolic extract of the seeds of *Trachysperum ammi* in Swiss albino mice. *Pharmacology & Pharmacy, 9*(12), 503.
66. Rastogi, M., Ojha, R. P., Prabu, P. C., Devi, B. P., Agrawal, A., & Dubey, G. P., (2012). Prevention of age-associated neurodegeneration and promotion of healthy brain ageing in female Wistar rats by long term use of bacosides. *Biogerontology, 13*(2), 183–195.
67. Sahreen, S., Khan, M. R., & Khan, R. A., (2014). Effects of *Carissa opaca* fruits extracts on oxidative pulmonary damages and fibrosis in rats. *BMC Complementary and Alternative Medicine, 14*(1), 1–9.
68. Sarker, J., Ali, M. R., Khan, M. A., Rahman, M. M., Hossain, A. S., & Alam, A. K., (2019). The plant *Aerva sanguinolenta*: A review on traditional uses, phytoconstituents and pharmacological activities. *Pharmacognosy Reviews, 13*(26), 89.

69. Sarris, J., Panossian, A., Schweitzer, I., Stough, C., & Scholey, A., (2011). Herbal medicine for depression, anxiety, and insomnia: A review of psychopharmacology and clinical evidence. *European Neuropsychopharmacology, 21*(12), 841–860.
70. Sekar, S., (2007). Traditional alcoholic beverages from ayurveda and their role on human health. *Indian Journal of Traditional Knowledge, 6*(1), 144–149.
71. Shakeri, A., Sahebkar, A., & Javadi, B., (2016). *Melissa officinalis* L.–A review of its traditional uses, phytochemistry and pharmacology. *Journal of Ethnopharmacology, 188*, 204–228.
72. Sharifi-Rad, J., Quispe, C., Vergara, C. V., Kitic, D., Kostic, M., Armstrong, L., & Cho, W. C., (2021). Genus viburnum: Therapeutic potentialities and agro-food-pharma applications. *Oxidative Medicine and Cellular Longevity, 2021*.
73. Shrestha, S., Park, J. H., Lee, D. Y., Cho, J. G., Cho, S., Yang, H. J., & Baek, N. I., (2012). Rhus parviflora and its biflavonoid constituent, rhusflavone, induce sleep through the positive allosteric modulation of GABAA-benzodiazepine receptors. *Journal of Ethnopharmacology, 142*(1), 213–220.
74. Srivastava, J. K., & Gupta, S., (2007). Antiproliferative and apoptotic effects of chamomile extract in various human cancer cells. *Journal of Agricultural and Food Chemistry, 55*(23), 9470–9478.
75. Srivastava, J. K., Shankar, E., & Gupta, S., (2010). Chamomile: A herbal medicine of the past with a bright future. *Molecular Medicine Reports, 3*(6), 895–901.
76. Suzuki, T., Pervin, M., Goto, S., Isemura, M., & Nakamura, Y., (2016). Beneficial effects of tea and the green tea catechin epigallocatechin-3-gallate on obesity. *Molecules, 21*(10), 1305.
77. Suzuki, T., Miyoshi, N., Hayakawa, S., Imai, S., Isemura, M., & Nakamura, Y., (2016). Beverage impacts on health and nutrition. *Beverage Impacts Heal. Nutr.*, 49–67.
78. Taiwo, A. E., Leite, F. B., Lucena, G. M., Barros, M., Silveira, D., Silva, M. V., & Ferreira, V. M., (2012). Anxiolytic and antidepressant-like effects of *Melissa officinalis* (lemon balm) extract in rats: Influence of administration and gender. *Indian Journal of Pharmacology, 44*(2), 189.
79. Tayab, M. A., Chowdhury, K. A. A., Jabed, M., Mohammed, T. S., Kamal, A. M., Islam, M. N., & Simal-Gandara, J., (2021). Antioxidant-rich *Woodfordia fruticosa* leaf extract alleviates depressive-like behaviors and impede hyperglycemia. *Plants, 10*(2), 287.
80. Teri, L., & Gallagher-Thompson, D., (1991). Cognitive-behavioral interventions for treatment of depression in Alzheimer's patients. *The Gerontologist, 31*(3), 413–416.
81. Unno, K., Fujitani, K., Takamori, N., Takabayashi, F., Maeda, K. I., Miyazaki, H., & Hoshino, M., (2011). Theanine intake improves the shortened lifespan, cognitive dysfunction and behavioral depression that are induced by chronic psychosocial stress in mice. *Free Radical Research, 45*(8), 966–974.
82. Valdés-Sustaita, B., López-Rubalcava, C., González-Trujano, M. E., García-Viguera, C., & Estrada-Camarena, E., (2017). Aqueous extract of pomegranate alone or in combination with citalopram produces antidepressant-like effects in an animal model of menopause: Participation of estrogen receptors. *International Journal of Molecular Sciences, 18*(12), 2643.

83. Wakabayashi, C., Numakawa, T., Ninomiya, M., Chiba, S., & Kunugi, H., (2012). Behavioral and molecular evidence for psychotropic effects in L-theanine. *Psychopharmacology*, *219*(4), 1099–1109.
84. Wang, H., Nair, M. G., Iezzoni, A. F., Strasburg, G. M., Booren, A. M., & Gray, J. I., (1997). Quantification and characterization of anthocyanins in Balaton tart cherries. *Journal of Agricultural and Food Chemistry*, *45*(7), 2556–2560.
85. Wang, H., Nair, M. G., Strasburg, G. M., Chang, Y. C., Booren, A. M., Gray, J. I., & DeWitt, D. L., (1999). Antioxidant and anti-inflammatory activities of anthocyanins and their aglycon, cyanidin, from tart cherries. *Journal of Natural Products*, *62*(2), 294–296.
86. Wang, L., Guo, T., Guo, Y., & Xu, Y., (2020). Asiaticoside produces an antidepressant-like effect in a chronic unpredictable mild stress model of depression in mice, involving reversion of inflammation and the PKA/pCREB/BDNF signaling pathway. *Molecular Medicine Reports*, *22*(3), 2364–2372.
87. Wu, L., Sun, D., & He, Y., (2016). Fruit and vegetables consumption and incident hypertension: Dose–response meta-analysis of prospective cohort studies. *Journal of Human Hypertension*, *30*(10), 573–580.
88. Yadav, C. K., Poudel, K., Mehta, R., & Shrivastava, A. K., (2020). Anti-depressant activity of the leaves of *Zanthoxylum Armatum* on Swiss albino mice. *Journal of Universal College of Medical Sciences*, *8*(2), 46–50.
89. Yang, C. S., Chen, G., & Wu, Q., (2014). Recent scientific studies of a traditional Chinese medicine, tea, on prevention of chronic diseases. *Journal of Traditional and Complementary Medicine*, *4*(1), 17–23.
90. Yin, C., Gou, L., Liu, Y., Yin, X., Zhang, L., Jia, G., & Zhuang, X., (2011). Antidepressant-like effects of l-theanine in the forced swim and tail suspension tests in mice. *Phytotherapy Research*, *25*(11), 1636–1639.
91. Zhu, W. L., Shi, H. S., Wei, Y. M., Wang, S. J., Sun, C. Y., Ding, Z. B., & Lu, L., (2012). Green tea polyphenols produce antidepressant-like effects in adult mice. *Pharmacological Research*, *65*(1), 74–80.
92. WHO, (2020). https://www.who.int/news-room/fact-sheets/detail/depression (accessed on 2 March 2024).

## CHAPTER 8

# Antidepression Agents in the Microalgae of the Himalaya

MUHAMMAD RIZWAN,[1] TAJ-UD-DIN,[1] NISAR UD DIN,[1] MOHAMMAD ALI,[1] NADIA BIBI,[2] ABDUR RAUF,[3] and HAFIZ ANSAR RASUL SULERIA[4]

[1]*Center for Biotechnology and Microbiology University of Swat, Swat, Tehsil, Charbagh, Khyber Pakhtunkhwa, Pakistan*

[2]*Department of Microbiology, Shaheed Benazir Bhutto Women University, Peshawar, Khyber Pakhtunkhwa, Pakistan*

[3]*Department of Chemistry, University of Swabi, Swabi, Anbar, Khyber Pakhtunkhwa, Pakistan*

[4]*School of Agriculture and Food, Faculty of Veterinary and Agricultural Sciences, The University of Melbourne, Parkville, Victoria, Australia*

## ABSTRACT

The promising anti-depressant properties revealed in preclinical and clinical trials are attracting numerous interests, despite the fact that the use of microalgae as a therapeutic strategy for depression is currently in its initial stages. Natural medicines with fewer side effects are often preferred over synthetic anti-depressant medications with potentially severe side effects. The previous reviews, as well as the first *in vitro* research of the anti-depressant effects of algal antioxidants, have given us a positive viewpoint on the potential use of microalgae as new anti-depressants in clinical research. Conceivably, approaching the usage of microalgae as an adjuvant

Himalayan Medicinal Plants for the Treatment of Depression: A Source of Rich Antidepressant Agents. Abdur Rauf, Prabhakar Semwal, & Hafiz Ansar Rasul Suleria (Eds.)

treatment in conjunction with traditional anti-depressants will inspire the creation of microalgae-based anti-depressants. Nonetheless, it is critical that these putative microalgae-based medicines meet the essential safety, quality, and effectiveness requirements set by drug regulatory regimes. An overview of the micro-algal biodiversity in the Himalayan area is given in this chapter. Utilization of microalgae as an anti-depressant, and their effects on society in relation to the main subjects, were also discussed.

## 8.1 INTRODUCTION

Although they belong to different categories of photosynthesizing microorganisms, microalgae and cyanobacteria are frequently discussed together [60]. In this context, this procedure will be utilized. In contrast to the prokaryotic cyanobacteria, which are more primitive, microalgae are complex eukaryotes [83]. It is estimated that there are between 45,000 and 100,000 species of algae. These species can be found in various habitats, including freshwater, saltwater, estuarine seas, deserts, ice sheets, soil, rocks, and tree bark [35, 44]. Generally, algae are considered to be a nuisance, causing problems in water parks, rivers, lakes, and oceans, as well as economic damage due to harmful algal blooms that lead to the repeated death of shellfish and marine animals [4]. However, algae are vital for life and have traditionally been a source of human survival [51]. For example, oceanic photosynthesis, primarily attributed to various algae, is responsible for producing approximately half of the oxygen in the atmosphere [8]. Microalgae and cyanobacteria are crucial for marine life and the production of a significant portion of our seafood. Seaweeds, also known as macroalgae, have been used by humans for at least 4,700 years as food, medicine, cosmetics, and fertilizer [21]. However, industrial production of macro- and microalgae is relatively recent [54]. All of our food and feed on land comes from plants that are descendants of algae. In fact, all species can be traced back to microalgae and their ancestors [97].

Several publications demonstrate the various ways in which microalgae have contributed to humanity [21, 29, 76]. Although they have considerable commercial potential, they play a far greater ecological significance in ensuring the survival of life on Earth. Commercial microalgae cultivation occurs in relatively small numbers and typically takes place in open ponds, frequently of the raceway variety [3]. Cultivation in totally closed photo-bioreactors is less common due to the high cost involved, and is usually employed for small-scale production of high-value algae [70]. Closed photo-bioreactors offer better contaminant control and growth conditions, allowing for a wider variety of microalgae to be cultivated compared to open ponds [34].

Currently, there are several pharmacological and psychotherapy options available for managing and treating depression [18]. Serotonin neurotransmitters, dopamine, serotonin inhibitors, and other second-generation antidepressant remedies are considered the first choice therapeutics for depression treatment due to their safety and tolerability compared to first-generation antidepressants (such as monoamine oxidase inhibitors and tricyclic antidepressants) [17, 95]. However, the use of second-generation antidepressants has been associated with various negative effects, such as drowsiness, abdominal discomfort (upset stomach, diarrhea, gastrointestinal hemorrhage, and acid reflux), scurvy, phytochemical and pharmacological investigations, and other adverse effects [36, 89]. Polyphenols derived from microalgae have been proven to have numerous medical benefits, particularly antioxidant potential [28]. The antidepressant properties of microalgae have opened up possibilities for the development of potential depression treatments. Considering the role of osmotic damage and neuroinflammation in the onset and progression of depressive episodes, microalgae could be a valuable resource for the production of cutting-edge antidepressant therapeutics [14]. This chapter has covered the medicinal importance of algal species from the Himalayan regions, especially for the treatment of depression.

## 8.2 DIVERSITY OF MICRO-ALGAE IN HIMALAYAN REGION ACCORDING TO DIFFERENT AGRO-CLIMATIC ZONES

### *8.2.1 DIVERSITY OF MICRO-ALGAE IN SHIVALIK HILL ZONE*

A wide variety of algae can be found in the Shivalik Hill Zone, which is located about 350 to 650 meters above sea level and has a subtropical environment. This zone includes the districts of Bilaspur, Una, Hamirpur, and portions of Sirmaur, Solan Kangra, and Chamba [81]. The algal biodiversity in various sites from this region has been investigated by many researchers, resulting in the discovery of numerous Phyla of algae species.

*Cyanophycean* flora has been reported from different districts of Himachal Pradesh. *Stichosiphon, Anabaenopsis, Aphanocapsa,* and

*Fortiea* were found in the Mandi, Una, and Hamirpur districts, respectively. *Cylindrocapsa* was exclusively found in the Sirmaur district. Pseudanabaena, *Cylindrospermum, Microcystis, Fortiea, Calothrix,* and *Anabaenopsis* were the key taxa studied. The following species were identified: *Oscillatoria, Lyngbya, Chroococcus, Aphanocapsa, Aphanothece, Merismopedia, Coelosphaerium, Gomphosphaeria, Stichosiphon, Anabaena, Nostoc,* and *Nodularia.* Around 27 of the 37 taxa were reported for the first time from various Himalayan locations [23, 26].

Sheer Khad, a tributary of the Sutlej River in the Western Himalayas, is important to the ecosystem of the Sutlej and the Gobind Sagar reservoir [5, 85]. Three locations on Sheer Khad were chosen for the study's sample collection (S1 to S3). Site S1 is located in the Jaroh area at 31°38'19.64" N and 76°42'59.09" E. Site S2 is located near Mundkhar (31°36'10.41" N, 76°41'38.27" E), district Hamirpur, on Sunail Khad. Sheer Khad's tributary Sunail Khad meets Sheer Khad near Jahu. Site S3 is located in the district of Bilaspur at (31°28'02.28" N, 76°41'25.29" E) Bhadrog. This study discovered 19 taxa, including *Bacillariophyta, Chlorophyta,* and *Cyanophyta/ Cyanoprokaryota.* The study's taxonomy included the following members of the phylum *Chlorophyta: Hydrodictyon reticulatum, Oedogonium* sp., *Pediastrum simplex, Rhizoclonium* sp., *Zygnema* sp., *Spirogyra* sp., and *Cosmarium* sp. The phylum Bacillariophyta contained *Synedra* sp., *Navicula viridula, Gomphonema* sp., *Nitzschia* sp., and Microcystis aeruginosa. The phylum *Cyanophyta* includes *Microcystis aeruginosa, Aphanothece castagnei, Anabaena* sp., and *Nostoc spongiaforme* [12, 81].

*Cholorophyceae* algae in the Kangra area of the Himalayan region was also reported [62, 81]. With a physical area of 5,739 km$^2$, the Himalayan Region's District Kangra is located in the Western Himalaya. The region's soils and water resources are remarkably diverse, offering significant potential for algal variety. Researchers investigated several unique algal species in Himachal Pradesh and collected samples from various districts. In total, they discovered 38 species of algae belonging to the *Chlorophyceae* phylum. This includes *Zygnema Gangeticum, Zygnema Gangeticum, Volvox Aureus, Tetraedron Minimum, Staurastrum Pseudotetracerum, Scenedesmus Bijugatus, Scenedesmus Arcuatus, Spirogyra Reinhardii, Spirogyra Regularis, Pediastrum Tetras, Spirogyra Punctiformis, Scenedesmus*, and *Quadricauda*, Green algae are still the least studied species in the state of Himachal Pradesh, despite its enormous importance [62, 81]. A thorough investigation and assessment of the *Chlorophycean* species present across the Himalayan area, as well as their proper documentation for preservation, cultivation, and economic exploitation, seem to be urgently needed.

### 8.2.2 DIVERSITY OF ALGAE IN MID HILL ZONE

Between 651 and 1,800 meters above mean sea level, this region has a warm, temperate climate. Parts of Mandi, Solan, Kullu, and Chamba, as well as Rampur in the Shimla district, Palampur in the Kangra Tehsil [55]. Researchers in this area have made admirable efforts to investigate numerous locations for new algal flora [81].

A study on the diversity of the *Xanthophyceae* and *Euglenophyceae* classes of algae from various regions in the district of Kangra was undertaken by Kumar et al. [48]. They also found the *Xanthophyceae* species *Botrydium granulatum, Vaucheria dichotoma*, and *Vaucheria sessilis*, as well as the *Euglenophyceae* species *Phacus longicauda* and *Phacus pleuronectes* [48, 81].

In order to investigate the dispersion pattern of cyanobacteria, researchers examined the algae varieties in the hot springs of the Himalayan region [61, 80]. From a structural, functional, adaptability, distribution, and economic standpoint, cyanobacteria are among the most fascinating types of algae. Researchers have identified various blue-green algae species in the Mandi area, including *Mastigocludus, Chroococcus, Lyngbya, Phormidium, Microcystis, Oscillatoria, Synechococcus, Fortiea,* and *Pseudanabaena* sp. [81].

Three freshwater lakes in District Mandi were researched by Thakur et al. in 2013 [96]. Prashar Lake and Kuntbhyog Lake were the subjects of the research in the Mandi district of Himachal Pradesh [96]. Over the course of two years, nine groups of algae with 138 species each were investigated. In the winter, common species including *Cymbella ventricosa* and *Cyclotella* sp. were discovered in Kuntbhyog Lake and Rewalsar. Both of these lakes also had abundant *Synedra* sp. The post-monsoon season saw a rise in *Cyanophytes*. According to their research, *Microcystis aeruginosa* was persistently prevalent throughout the years in Kuntbhyog Lake. *Spirulina gomontii*, which grows at its highest rate at Kuntbhyog Lakes, was one of several *Spirulina* species that were frequently seen there. In Rewalsar and Kuntbhyog Lakes, *Oscillatoria stigonema, Oscillatoria princeps* and *Oscillatoria limosa* were discovered to be prevalent throughout the seasons. The lush development of *C. vulgaris* was also observed in Kuntbhyog Lake. During the monsoon and post-monsoon seasons, *Volvocales* such as *Chlorogonium* sp. and *Chlamydomonas reinhardi* were found, with *C. reinhardi* being more abundant. *Cryptomonas erosa,* the sole species of *Cryptophyceae*, was observed during the investigation. *Synura adamsii* and Dinobryon species were detected at the Prashar site during the winter months, but not at the other two locations. *Ceratium hirundinella* and *Peridinium inconspicuum* are dinophycean members. All five *Euglena* species found at the Kuntbhyog and Rewalsar locations followed a similar pattern of presence and seasonality [81, 96].

Jindal and Thakur conducted research on the plankton diversity of Kuntbhyog Lake [96]. Situated at an elevation of 1,750 meters above mean sea level, Kuntbhyog Lake is located at the picturesque hilltop of Rewalsar town (31°37' N, 76°49'6" E), surrounded by hills on both sides. It is situated 12 kilometers from Rewalsar town in the northern Mandi district in the Himalayan area. The majority of the plankton in the lake consisted of phytoplankton. *Bacillariophyceae* was the dominant component, comprising 55.98% (29.28–77.82%), followed by *Chlorophyceae* at 27.68% (17.19–36.10%). *Cyanophyceae* accounted for 13.49% (0.0–35.03) and *Euglenophyceae* accounted for 2.85% (0.0–13.32%). *Bacillariophyceae* were abundant from September through October and February through March, while *Chlorophyceae* dominated in March and April, and *Cyanophyceae* and *Euglenophyceae* in May and June. A total of 48 species of plankton were identified, including *Oscillatoria* sp., *Cosmarium* sp., *Navicula* sp., *Scenedesmus* sp., *Closterium* sp., *Cymbella* sp., and *Nitzschia* sp. among the phytoplankton, and *Cylops* sp., *Keratella tropica*, and *Brachionus* sp. among the zooplankton. During the summer and after the monsoon, there was a profusion of plankton. The study found that high turbidity, gloomy weather, and salt concentration dispersion during the monsoon were associated with lower plankton quantity, while higher temperatures, light intensity, rigidity, pH, and nutritional content were advantageous for plankton abundance [81, 96].

According to Jindal et al., a shallow eutrophic, natural mid-altitude in the Himalayan area, located on a mountain spur in the Mandi district of Himachal Pradesh, India, has been the site of studies on phytoplankton dynamics and species diversity [43]. From these waters, 47 species have been recognized. A perpetual habit was shown by *Microcystis aeruginosa* and *Synedra ulna.Ankistrodesmus falcatus*, *Spirulina gomonti, Euglena oxyuris, Euglena acus, Synedra ulna, Navicula cuspidate, Eudorina elegans, Chlamydomonas reinhardi, Scenedesmus bijugatus, Chlorella vulgaris, Microcystis aeruginosa* and *Oscillatoria limosa* were extremely numerous [43].

Based on the same pattern diversity, Jindal et al. investigated the water quality and phytoplankton dynamics of Prashar Lake in Himachal Pradesh, India [43]. Prashar Lake is situated in the Mandi district of Himachal Pradesh, India (longitude: 76_3702000-77 2301500 E; latitude: 31_1305000-32 0403000 N). A total of 67 phytoplankton species from eight different groups of algae were found. The most diversified groupings included *Volvocales, Cyanophyceae, Zygnematales, Chlorococcales,* and *Bacillariophyceae*. The biggest number of species, 31 species, were found in the family *Chlorophyceae*, which also included members of the *Volvocales, Chlorococcales, Cladophorales,* and *Zygnematales* [43]. Among the *Bacillariophyceae*, the main algae species were *Pinnularia gibba, Navicula* sp., *Gomphonema gracile, Cocconeis* sp. and *Fragilaria capucina* [43].

Renuka Wetland's productivity and trophic status were investigated by Jindal and Prajapat [42] (dist. Sirmour). Located in Sirmour District, Renuka Wetland is situated 173 kilometers southwest of Shimla at an elevation of 645 meters above sea level. Their research revealed that phytoplankton (algae) constituted the majority of the total plankton. The main component of phytoplankton was *Bacillariophyceae*, whereas *Chlorophyceae* and *Cyanophyceae* were subdominant. *Bacillariophyceae* were abundant in October-November and March, whereas *Cyanophyceae* and *Eugelophyceae* were abundant in May-June. During March and April, the *Chlorophyceae* were plentiful. *Ankistrodesmus, Closterium, Cymbella, Euglena, Fragilaria, Microcystis, Navicula, Nitzschia, Oscillatoria,* and *Scendesmus* were among the phytoplankton species discovered [42].

### 8.2.3 DIVERSITY OF ALGAE IN COLD DRY ZONE

This region includes the districts of Lahaul-Spiti and Kinnaur, as well as Pangi Tehsil in the Chamba District, and is located around 1,800 meters above mean sea level [77]. Four species of *Lagerheimia Chodet* algae were identified from Nako Lake [26, 81]. The lake is situated at an elevation of 3,662 meters above sea level in the Pooh subdivision of the Kinnaur district of the Indian Himalayan region, which is part of the Indo-western

Himalaya [45]. *Lagerheimia Chodat, L. ciliata* (*Lagerheim*) *Chodat, L. subsalsa, Lemmernann, L. citriformis* (Snow) *Collins*, and *L. chodatii* were among the freshwater algal genus [26, 81]. Except for *L. chodatii* all of these species are new to the Indian freshwater algal flora.

Singh et al. studied the diversification of cyanobacteria in high elevation lakes in Lahul Spiti, Himachal Pradesh [87]. The four lakes investigated were Chandra Tal, Deepak Tal, Sissu Lake, and Suraj Tal. These lakes are located in the Lahaul-Spiti cold desert area of Himachal Pradesh's northern Himalayas. *Synechocystis pevalekii,Cyanobium parvum, P. chalybeum, Phormidium autumnale, L. valderiana Limnothrix redekii, L. frigida, Nodularia sphaerocarpa, Nostoc linckia,* and *Planktothrix agardhii* were among the algal species isolated from this region [87].

## 8.3 MICRO-ALGAE AS A HUMAN FOOD

Despite its significant protein content, dried microalgae have not gained widespread acceptance as a meal or dietary alternative [7]. Although microalgae are used in the commercial production of food spices (such as pigments and artificial colors) and nutritional supplements (such as Provitamin A or β-carotene) [22], they are not widely consumed as food. Due to its low digestibility, *Chlorella* microalgae is added to various meals but cannot be used as a primary nutrient source [22]. Numerous cosmetic products and anti-aging cosmetics, based on seaweed and microalgae extracts, have been introduced [74].

In addition to seaweed, *Spirulina* has a long history as a specialized nutrition [2, 58]. Native people of America, Africa, and Asia have been consuming *Arthrospira* for almost 1,000 years [16]. The Mexicans used to harvest Arthrospira, also known as "tequitlatal" or "tecuitlatl," from nearby lakes for nutrition [67]. Until the 16th century, biomass was collected using fine nets. Consumption continued for over 100 years after the Spanish invasion but has largely disappeared as the region's lagoons and lakes were filled for construction [2].

*Arthrospira* cultivates naturally in shallow alkaline water reservoirs near Lake Chad's sandy northeastern shoreline in Chad, West Africa[84]. Pool water is drained through cloth to concentrate the biomass slurry. The traditional method involves pouring the condensed slurry onto a spherical piece of sand to absorb water. After 20 minutes, the thin wet cake adhered to the sand is sliced into squares, removed, and further dried [84]. Local markets sell this thin sun-dried *Arthrospira* cake, known as "dihe." Dihe is produced year-round by indigenous Kanembu women and is often served as a sauce. The dried cakes are crushed or pulverized before being combined with water and other ingredients, and then cooked [68].

The sauce is accompanied by a thick bread composed of cooked millet or sorghum. Every year, about 250 metric tons of dried *spirulina* are consumed as "dihe" [47]. The global commercial output of *Arthrospira* biomass is expected to be approximately 130 to 150 tons of dry biomass [72]. As a result, production in Chad's Kanem area accounts for over 10% of global output. The nutritional features of *Arthrospira* and its modern commercial production are extensively described in the literature [9, 10, 37, 50, 82]. *Arthrospira* is manufactured by machines in China and the United States. Simple manufacturing methods are being adopted in underdeveloped locations to meet limited demands. *Arthrospira*-based culinary items may contain colorants like C-phycocyanin and extracellular biopolymers. Some microalgal oils and proteins may have dietary value. Not all microalgae are edible, and some are harmful [50].

*Nostoc* is a terrestrial colonial cyanobacterium with a troubled antiquity of social use as nutrition, notably in China [31]. *Nostoc*, often known as "hair vegetable," has been used in Chinese medicine and specialty dishes for over 1,000 years [30]. Consumption occurs in salads and soups in various other Asian nations. Although the evidence for *Nostoc* being a health food with therapeutic effects is unclear, it is believed to have been obtained entirely in the wild, with no industrial production. *Nostoc* was wild harvested in China by scraping it from the soils of the dry and Northwest drylands and Inner Mongolia. Harvesting has been prohibited due to significant soil erosion and desertification caused by indiscriminate harvesting. Many of the fat choy items on the market today are forgeries made of colored starch noodles. Other colonial *Cyanobacteria,* including *Nostoc,* are eaten in China and Taiwan [21, 56].

## 8.4 MICROALGAE AND SEAFOOD

In most aquatic ecosystems, microalgae and cyanobacteria are the primary producers, constituting the main source of seafood [66]. Global fish consumption has more than doubled since 1961, reaching an average of 20 kilograms per person per year in 2018 [21]. Additionally, the consumption of seafood (including freshwater species) has also increased [21].

However, due to the current state of worldwide fisheries, further increases in wild fish catches are unlikely [11].

Fish is a valuable source of protein in many regions. Humans consume over 75% of the world's fish catch, with the remainder being processed into fishmeal and oil for feeding farmed animals and fish [11, 21]. The production of seafood through farming is known as aquaculture [75]. Despite fish farming having a history of over 3,500 years, significant aquaculture is a relatively recent phenomenon [38]. As wild fish catches approach their limits, aquaculture now accounts for approximately half of all seafood consumed and is projected to replace wild-caught supplies. Aquaculture production currently exceeds global cattle productivity [21].

The issue is that aquaculture is heavily reliant on marine wild capture since the feed required contains fishmeal and oil. In order to produce 1 kilogram of live weight of aquacultured fish, normally 0.3 kilograms of fresh wildly caught fish is required, in addition to other feed components [21, 75]. Approximately 75% of the total dietary protein comes from marine fish, primarily obtained for aquaculture production as well as bycatch from trawlers. Sardines, anchovies, menhaden, and sand eels are among the fish used for their proteins and oil. A smaller amount (25%) of aquatic proteins and oil for human use is obtained from byproducts of fish processing [21, 75]. The top 10 producing nations generated an average of 679,000 million tons of fish oil per year between 2012 and 2015. Data for annual worldwide output of fish oil from 2007 to 2016 show flat or declining output, possibly due to a scarcity of wild-caught fish [21].

Wild harvesting of fish cannot be used to increase the supply of fish oil and meal for aquaculture in a sustainable manner. In this context, micro-algae are expected to play a larger role in the future, as algal meal and oil, when combined with animal and plant protein, have the potential to replace or significantly reduce aquaculture's reliance on fish oil and meal. Soybeans and other cereals currently provide some of the nutrients needed in aquaculture feeding, but they cannot fully substitute fish meal and oil. Crops lack certain essential nutrients found in fish meal and oil [21, 75].

Plant proteins, for example, may be undersupplied with amino acids such as tryptophan, threonine, methionine, and lysine. Many microalgae are rich in all the necessary amino acids. Another essential nutrient provided by microalgae but deficient in plants is the non-protein sulfonic acid taurine, which is obligatory for numerous predatory fish. Moreover, all of the critical long-chain polyunsaturated fatty acids (PUFA) contained in fish oil are produced by microalgae. These fatty acids are critical for the health of marine animals. Some marine species, for example, crabs, lobsters, red mullet, red snapper, salmon, and shrimp, require bio-accumulated carotenoids originating from microalgae, such as astaxanthin, which can be obtained from cultivated microalgae. Theoretically, treated microalgal biomass could be a substitute for fish meal and oil as feeds in aquaculture if generated cheaply and in sufficient quantities [21, 75].

Since the use of microalgal species as feedstuffs in aquaculture is well recognized, feeder microalgae are expensive due to the low production of photoautotrophic growth. This low production is due to a failure to provide adequate light to self-shaded microalgal culture, not to any fundamental constraint in algal biology. Microalgae grown heterotrophically on organic carbon derived from higher plants have substantially higher biomass production (e.g., 10-fold higher) than photoautotrophically developed cultures. As a result, heterotrophic growth may have a better chance of meeting the expected need for PUFA-rich algal feeding for aquaculture, which could be used to replace fish oils [21, 75].

Oils rich in PUFAs may now be produced in genetically modified (GM) plants, posing a challenge to the increasing use of microalgae biomass as aquafeeds. Theoretically, it is possible to completely replace fish meal and oil in aquaculture diets with a combination of plant and microbial nutrients and harvested oil. Similarly, genetically engineered microalgae for aquaculture feedstuff may be used to overproduce the necessary PUFAs [21, 75].

## 8.5 PROBABLE BIOFUELS AND SUPPLEMENTARY ORGANIC CHEMICALS

Photo-synthetically generated microalgae can synthesize a diverse range of organic compounds, particularly biomass for a variety of biofuels, under the correct circumstances. In fact, the energy contained in these compounds is obtained from the sun through photosynthesis. The starch that algae can store in large quantities is formed directly from the glucose produced during photosynthesis. Glucose is produced by the straightforward hydrolysis of starch, and this glucose can then be fermented to produce fuels like bioethanol and biobutanol. Similar to how vegetable oils are converted into biodiesel by transesterification using methanol, algae-derived triglyceride oils can also be readily converted into this fuel. Algal biomass can be digested anaerobically to produce biogas, a combustible mixture of methane and carbon dioxide. Additionally, the biomass can be fermented to create high-energy hydrogen. Additionally, hydrogen can be produced directly from algae. Techniques like pyrolysis and hydrothermal conversion can be used to treat algae biomass to form biofuel [20, 52, 92].

A bio-refinery can manufacture various industrial chemicals from algal biomass or other biomass in the same way as a petroleum refinery can from crude petroleum and coal. Although commercial bio-refineries exist, the feedstock used is mostly lignocellulose biomass rather than algal biomass. In theory, algal biomass can be utilized for this purpose, but it is too costly. Structured and bioactive proteins, tiny organics and sterols, pigments, and colorants, polysaccharides, bioplastics, and ultraviolet (UV) light screening agents are some more algal products and prospective products. Other items can be manufactured [20, 25, 52, 92].

## 8.6 CARBON SEQUESTRATION BY MICROALGAE

It is a common misconception that the Earth's atmosphere contained plenty of gases like methane and carbon monoxide before photosynthesis began to evolve oxygen. Additionally, the number of rocks in the Earth's crust decreased [27, 69, 86]. Rock oxidation arose after the initial appearance of oxygenic photosynthesis because the oxygen it produced changed methane and carbon monoxide into carbon dioxide. The Great Oxygenation Event (GOE), also known as the Oxygen Catastrophe, occurred when these oxygen sinks ran out of oxygen and dioxygen ($O_2$) started to build up in the atmosphere. The concentration of carbon dioxide in the atmosphere progressively decreased due to the photosynthesis of plants, cyanobacteria, and algae [27, 69, 86].

About half of the dry biomass of algae is made up of carbon. Almost all of this carbon is derived from atmospheric $CO_2$ through photosynthesis. The decomposition of dead algal biomass resulted in the release of most of the absorbed carbon back into the atmosphere. In anoxic settings like marshes, dead woody plant biomass occasionally did not fully decompose [27, 69, 86]. Under the right circumstances, some of this partially decomposed material that had accumulated as peat was able to be converted into coal and natural gas over the course of geologic time. A similar process produced fuel from the remains of extinct cyanobacteria and microalgae. Carbon was slowly trapped in coal and other fuel formations over many centuries [27, 69, 86].

Furthermore, the carbon is also currently being fixed. Net world primary manufacturing, which measures the quantity of carbon fixed per annum through photosynthesis, surpasses 100 billion tons per annum. Algae and cyanobacteria in the waters fix around half of this [40]. Since a large portion of the carbon is released during the breakdown of dead algal and cyanobacterial biomass, in most cases, this carbon reduction is not long-lasting [27, 69, 86].

Human activity began a rapid and continuous release of carbon that had been trapped for millions of years, primarily during the 1750s. For instance, in 2016 alone, over 5.6 billion metric tons of carbon were emitted into the atmosphere solely from coal combustion. The rapid economic development over the past 220 years has been fueled by the release of carbon that had been trapped for millennia. Consequently, the industrial revolution, which began in the late 1700s, resulted in the rapid release of fossilized carbon sources such as petroleum and other fuels, leading to an increase in the atmospheric concentration of carbon dioxide [27, 69, 86].

Since the 1750s, the amount of carbon dioxide in the atmosphere has increased by more than 40%. The oceans absorb between 30% and 40% of the carbon dioxide emitted by human activities, resulting in ocean acidification. Deforestation, by releasing carbon from plants and soils, further contributes to the accumulation of carbon dioxide in the atmosphere [27, 69, 86].

Algae not only sequester carbon through the creation of peat, coal, and petroleum, but they also store carbon by forming inorganic deposits. The white cliffs of Dover, England, for example, were formed in this way. Chalk, which is made of calcium carbonate, is used to create these cliffs. The majority of the chalk rock and limestone deposits were produced by the deposition of the calcium carbonate skeletal remains of coccolithophore microalgae. Coccolithophores are still the main producers of calcite in the oceans, manufacturing calcite rocks. On average, it is predicted that coccolithophores deposit 1.4 billion metric tons of calcite on the ocean floor each year. These calcite deposits act as a significant carbon sink, continuously removing carbon dioxide from the atmosphere [27, 69, 86].

## 8.7 HARMFUL MICROALGAE

Some phytoplankton and microalgae are harmful to humans and have a negative impact on the economy. One example of this is red tides, which are toxic algal blooms. These blooms have been a problem for society since ancient times. Harmful algal blooms caused by toxic dinoflagellate microalgae are detrimental to aquatic life. Consuming contaminated seafood can cause harm to people. Both red tides and green blooms can kill marine life, including fish, birds, and other sea creatures. They can also cause illness and death in humans, as well as economic instability. Algal blooms are a natural occurrence, but they have become more frequent in some areas, possibly due to nutrient releases from human activities and the unintentional transfer of algal cysts in ship ballast water. The toxic algae may actually be a source of medicines, so they are not entirely harmful [53, 57].

## 8.8 COMMERCIAL MANUFACTURING AND PROCESSING OF MICROALGAL BIOMASS

In addition to the properties of microalgae biotechnology described earlier in this chapter, several other considerations need to be taken into account when developing microalgae-based industrial processes. Algal physiology and metabolism, which impact metabolite production, are some of the fundamental factors that must be considered. Special care is required for the preservation, conservation, and reproduction of the algal strain, as the loss of a high-producing strain could jeopardize the financial feasibility of the entire commercial enterprise. Synthetic biology methods can be used to produce new chemicals and goods in microalgae. When a significant volume of biomass has been produced in open ponds or photo-bioreactors, it is necessary to collect or separate the biomass from the water. If the desired outcome is an intracellular metabolite rather than the algal biomass itself, it must be extracted in some manner, either immediately from the cells or after they have been vigorously disrupted [1, 13].

The establishment of commercial operations requires careful scaling up and a combination of several processing steps in order to achieve optimal results. To develop a profitable manufacturing strategy, options for process intensification must be assessed. The product life cycle should be evaluated for ethical and legal reasons to minimize environmental impact. Additionally, energy-saving practices are necessary to address financial and environmental factors [1, 32].

## 8.9 MICRO-ALGAE AS AN EMERGENCE THERAPEUTIC FOR DEPRESSION

About 50% of individuals with depression have been reported to utilize complementary and alternative treatments. Microalgae have been utilized since ancient times as feed, fertilizer, and a source of therapeutics. More recently, they have also been used as a raw material in the production of nutrition, beauty products, and medications. There are various types of microalgae found in the ocean, including diatoms, dinoflagellates, green algae, and blue-green algae (cyanobacteria). Marine microalgae are suitable for use in the nutraceutical and pharmaceutical industries due to their fast growth and ability to thrive without freshwater or agricultural land. Furthermore, marine algae possess abundant secondary metabolites with therapeutic properties that are not found in plant species. Some of these compounds, such as alginates, carotenoids, fucoidan, phlorotannin, sargaquinoic acids, and sulfated polysaccharides, exhibit anticoagulant, antimicrobial, antioxidant, anti-allergic, anti-cancer, anti-inflammatory, anti-obesity, and neurological activities [71, 91].

### *8.9.1 PRE-CLINICAL STUDIES*

#### *8.9.1.1 ARTHROSPIRA PLATENSIS*

*Arthrospira platensis* is indeed a real species of blue-green, spiral-shaped, freshwater filamentous microalga. It was originally known as *Spirulina platensi* and is more often referred to as *Spirulina. A. platensis* extensively isolated in subtropical alkaline lakes, East African soda lakes, and Chenghai Lake in China [24, 39]. Its dried biomass has been promoted as superfoods full of vitamins, minerals, proteins, carbohydrates, and PUFAs [19, 49]. *Spirulina* showed an anti-depressant impact in the forced swim test (FST) on a mouse model of depression, according to research done by Suresh et al. [93]. It was noted that ingesting malted barley-derived hydrolyzed *spirulina* extract at a dose of 10 ml per kg body weight per day for two weeks potentially reduced signs and symptoms of depression [93]. A reduction in lactate dehydrogenase release was associated with a reduction in the immobility time in FST. Conversely, there is no impact of hydrolyzed extract on the levels of albumin, glucose, total proteins, and creatine kinase [46].

In a research study, *A. platensis* was found to exhibit anti-depressant effects in rat and mouse depression models using various tests such as the tail suspension test (TST), the FST, aggressive behavior induced by clonidine, aggressive behavior elicited by levodopa (L-DOPA), and induced 5-HT head twitches [91]. Administration of *A. platensis* orally at a dosage of 100, 200, or 400 mg per kg of body weight per day for 7 days resulted in a decrease in depression symptoms and potentially reduced the time animals spent in stasis during TST and FST. These tests were conducted one hour after oral administration of *A. platensis* or injection of common medicines (15 mg per kg imipramine or 2.5 mg per kg lorazepam), except for the FST on day 8. In the case of depression, *S. platensis* was found to improve neurotransmission [91]. The rate of induced 5-HTP head twitches, aggressiveness caused by clonidine, hyperactivity induced by L-DOPA, and violent behavior were all significantly increased, suggesting that *S. platensis* may enhance dopaminergic, noradrenergic, and serotonergic pathways in depression [91].

Furthermore, pretreatment with *S. platensis* potentially improved catalase levels while reducing the rate of lipid peroxidation, indicating the plant's strong antioxidant capability [91]. The pathophysiology of depression is characterized by an increase in lipid peroxidation products and DNA damage, which affects the viscosity and phospholipid content of neuronal membranes, leading to their degradation. These events disrupt catecholaminergic and serotonergic neurotransmission. Malondialdehyde also hinders serotonin from binding to receptors, disrupting the delicate balance between oxidative stress and serotonin metabolism. Therefore, the antioxidant property of *A. platensis* contributes to its anti-depressive properties [91].

According to Moradi-Kor et al., *S. platensis* prevented the basolateral amygdala of adult female rats subjected to restraint stress from undergoing molecular and morphological remodeling. Adolescent stress decreased apical dendritic length and pyramidal neuron branch points while increasing oxidative stress measurements [59]. Administration of *S. platensis* orally at a dosage of 200 mg per kg of body weight per day for 15 days decreased oxidative stress and improved ferric-reducing antioxidant capacity (FRAP) [90, 91].

It has been shown that stressful prepubertal experiences alter how the body reacts to stress and make people more susceptible to stress-related disorders as adults [6]. The prefrontal cortex, hippocampus, and amygdala may undergo considerable remodeling as a result of ongoing stress, which may result in mood disorders including anxiety and depression [6]. Intriguingly, Moradi-Kor et al. found that teenage stress decreased the basolateral amygdala's expression of tropomyosin receptor kinase B (TrkB) and brain-derived neurotrophic factor (BDNF) [59].

#### 8.9.1.2 BOTRYOCOCCUS BRAUNII

Triterpene-containing *Botryococcus braunii* is a green colonial microalga with a pyramidal shape. It can be found worldwide in freshwater reservoirs, ponds, and lakes [98]. In the FST-induced mouse model of depression, Sasaki et al. discovered that *B. braunii* ethanol extract exhibits anti-depressant properties [78]. The *B. braunii* ethanol extract was administered orally daily for 14 days at a dosage of 100 mg/kg body weight. This treatment alleviated depression-like behaviors and reduced immobility in the FST [78]. The synthesis of dopamine (arginine/serine-rich coiled-coil 1, Rsrc1), energy metabolism (polyribonucleotide nucleotidyl-transferase 1, Pnpt1), and neurogenesis were all increased. The expressions of mRNAs for pyruvate carboxylase (PC), 3-monooxygenase (TH), tyrosine, and BDNF were also elevated. The authors hypothesized that the high energy output resulted from the regulation of neurogenesis and an increase in dopaminergic activity, indicating effects on stress and depression [78].

#### 8.9.1.3 CHLORELLA VULGARIS

A freshwater, single-celled, greenish microalga with a high concentration of chlorophyll is known as *Chlorella vulgaris* [65, 73]. It has been shown that *C. vulgaris* possesses pharmacological benefits such as anti-inflammatory, antioxidant, and anti-aging properties [65]. Soetantyo and Sarto investigated the effects of commercially produced and farmed *C. vulgaris* on rats' depression induced by chronic unpredictable mild stress (CUMS) [88]. After a 42-day period of stress induction, a 14-day oral treatment of *C. vulgaris* at a dosage of 360 mg/kg body weight/day alleviated depression symptoms, reduced immobility time in the forced swim test (FST), and restored sucrose preference in the sucrose preference test (SPT). These findings indicate that *C. vulgaris* has neuroprotective effects against CUMS [88].

#### 8.9.1.4 HAEMATOCOCCUS PLUVIALIS

*Haematococcus pluvialis* is found in temperate areas all over the world[79]. Due to its ability to collect significant amounts of astaxanthin. A red keto-carotenoid pigment, it has commercial potential [33]. The properties of AST on LPS-induced neuro-inflammation and comparable depressive behaviors were investigated in a depressive mouse model [63]. The induced immobility length by LPS in TST and FTS was reversed by daily orally administered AST at different doses (mg per kg of body weight) for 7 days, followed by an intraperitoneal injection of LPS. This resulted in a reduction in depressive-like behaviors [63].

The antidepressant and neuroprotective properties of AST were mediated through the restoration of pro-inflammatory cytokines, inflammatory mediators, COX-2 mRNA expression, and nuclear factor kappa B (NF-κB) in the prefrontal cortex and hippocampus. Additionally, the advantages of AST's neuroprotection may be due to its ability to integrate itself within the phospholipid bilayer or on the surface of the membrane [41, 63].

Similarly, Jiang et al. demonstrated that the effects of AST are comparable to those of antidepressants in a model of adult depressive mice [41]. Administration of various oral doses of astaxanthin one hour prior to neurochemical and behavioral testing potentially reduced immobility time in the TST and FST, as well as improved depressive-like behaviors, but had no impact on locomotion. The anti-immobility effect of AST in the FST and TST was abolished after pretreatment with para-chlorophenylalanine, suggesting the involvement of the serotonergic system in the antidepressant-like effects of AST. Acute AST treatment also increased noradrenaline levels in the hippocampus, but had minimal effect on the hypothermia and ptosis induced by reserpine. Furthermore, astaxanthin lowered the activity of indoleamine 2,3-dioxygenase (IDO) in the hypothalamus, cortex, and hippocampus. This led to a decrease in the tryptophan/KYN ratio, while the tryptophan/serotonin ratio increased. Therefore, the antidepressant properties of AST are facilitated through the central serotonergic system [41].

## 8.10 CLINICAL STUDIES

### *8.10.1 CHLORELLA VULGARIS*

In addition to standard medication, Panahi et al. investigated the therapeutic efficacy of an antioxidant-rich *C. vulgaris* extract (ALGOMED®) in patients with severe depression [64]. The Diagnostic and Statistical Manual (DSM) of Mental Disorders-IV was used to diagnose 92 people with major depression for the study (DSM-IV). The patients received standard antidepressant care. Around 50 people received standard antidepressant medication, while 42 others received CVE as an additional therapy. For six weeks, participants in the CVE intervention group took six (ALGOMED®) 300 mg tablets daily, totaling 1,800 mg of CVE. According to results from the Beck Sadness Inventory II (BDI-II) and other measures, after oral administration of CVE for 6 weeks, the intervention group showed a decrease in affective signs of hopelessness and depression, but not physical or cognitive symptoms [64].

### *8.10.2 HAEMATOCOCCUS PLUVIALIS*

The effects of astaxanthin derived from *H. pluvialis* on depression and fatigue were studied by Talbott et al. in 28 individuals [94]. The study included adults without depression who were in good health, active, and recruited for the study in a 1:1 ratio of males to females. Over the course of eight weeks, participants in the *H. pluvialis* intervention group were administered a daily dose of 12 mg of astaxanthin (AstaZine®), while those in the control group received a matching placebo. The intake of astaxanthin, as evaluated through the profile of mood states (POMS) questionnaire, resulted in a significant reduction in fatigue and depressive symptoms. The POMS includes six mood subscales: anxiety, depression, anger, vigorous activities, fatigue, and confused perplexity [94].

## 8.11 LIMITATIONS AND FUTURE DIRECTIONS

In addition to being consumed frequently, microalgae may possess antidepressant qualities, as suggested by animal studies and anecdotal clinical experiences. Well-designed randomized controlled trials are considered the gold standard for determining the safe clinical therapeutic benefits of natural substances, and should not be substituted for such findings.

Strong scientific evidence is deemed necessary for randomized clinical trials, including single-blind or double-blind placebo-controlled studies, to support the efficacy of natural compounds examined in preclinical research [15, 99]. The majority of clinical findings used in the current study were observational. Therefore, randomized studies are still required to establish the effectiveness of microalgae in inducing anxiolytic and depressive-like effects before considering routine therapeutic usage for patients with depression.

## 8.12 SUMMARY

The promising anti-depressant properties revealed in preclinical and clinical trials are attracting a lot of interest, despite the fact that the use of microalgae as a therapeutic strategy for depression is currently in its initial stages. Synthetic antidepressants with potentially serious adverse effects are frequently preferred over natural drugs with fewer side effects. We have a positive attitude towards the potential use of microalgae as novel anti-depressants in clinical practice, thanks to the reviewed research and the first *in vitro* assessment of their anti-depressant properties. Perhaps in the near future, the development of microalgae-based anti-depressants will be stimulated by the use of microalgae as an adjuvant therapy alongside conventional anti-depressants. However, it is imperative that these purported microalgae-based medications satisfy the fundamental safety, quality, and efficacy standards established by drug regulatory systems.

## KEYWORDS

- **algae**
- **anti-depressant**
- **depression**
- **Himalayan**
- **Himalayan region**
- **microalgae**
- **natural medicines**
- **photosynthesizing microorganisms**

## REFERENCES


1. Acién, F., Fernández, J., & Molina-Grima, E., (2014). Economics of microalgae biomass production. *Biofuels from Algae* (pp. 313–325). Elsevier.
2. Ali, S. K., & Saleh, A. M., (2012). Spirulina-an overview. *International Journal of Pharmacy and Pharmaceutical Sciences, 4,* 9–15.
3. Amaro, H. M., Guedes, A. C., & Malcata, F. X., (2011). Advances and perspectives in using microalgae to produce biodiesel. *Applied Energy, 88*, 3402–3410.
4. Anderson, D. M., Fensin, E., Gobler, C. J., Hoeglund, A. E., Hubbard, K. A., Kulis, D. M., Landsberg, J. H., et al., (2021). Marine harmful algal blooms (HABs) in the United States: History, current status and future trends. *Harmful Algae, 102*, 101975.
5. Bain, W. K., (2018). Some preliminary observations on the Acheulian lithic remains of Bam locality of Indian Siwalik Hills, Himachal Pradesh. *Journal of the Anthropological Survey of India, 66*, 283–307.
6. Bazak, N., Kozlovsky, N., Kaplan, Z., Matar, M., Golan, H., Zohar, J., Richter-Levin, G., & Cohen, H., (2009). Pre-pubertal stress exposure affects adult behavioral response in association with changes in circulating corticosterone and brain-derived neurotrophic factor. *Psychoneuroendocrinology, 34*, 844–858.
7. Becker, E. W., (2007). Micro-algae as a source of protein. *Biotechnology Advances, 25*, 207–210.
8. Beer, S., Björk, M., & Beardall, J., (2014). *Photosynthesis in the Marine Environment.* John Wiley & Sons.
9. Belay, A., (1997). Mass culture of spirulina outdoors—The earthrise farms experience. *Spirulina Platensis*, 131–158.
10. Belay, A., (2007). Spirulina (*Arthrospira*): Production and quality assurance. *Spirulina in Human Nutrition and Health* (pp. 15–40). CRC Press.
11. Bell, J. D., Watson, R. A., & Ye, Y., (2017). Global fishing capacity and fishing effort from 1950 to 2012. *Fish and Fisheries, 18*, 489–505.
12. Bhakta, S., Das, S., Nayak, M., Jena, J., Panda, P., & Sukla, L., (2011). Phyco-diversity assessment of Bahuda river mouth areas of east coast of Odisha, India. *Recent Research in Science and Technology, 3.*
13. Bhalamurugan, G. L., Valerie, O., & Mark, L., (2018). Valuable bioproducts obtained from microalgal biomass and their commercial applications: A review. *Environmental Engineering Research, 23,* 229–241.
14. Brahmachari, G., (2013). *Chemistry and Pharmacology of Naturally Occurring Bioactive Compounds.* CRC Press Boca Raton, FL, USA.
15. Calixto, J., (2000). Efficacy, safety, quality control, marketing, and regulatory guidelines for herbal medicines (phytotherapeutic agents). *Brazilian Journal of Medical and Biological Research, 33*, 179–189.
16. Capelli, B., & Cysewski, G. R., (2010). Potential health benefits of spirulina microalgae. *Nutrafoods, 9*, 19–26.
17. Carek, P. J., Laibstain, S. E., & Carek, S. M., (2011). Exercise for the treatment of depression and anxiety. *The International Journal of Psychiatry in Medicine, 41*, 15–28.
18. Carvalho, A. F., Berk, M., Hyphantis, T. N., & McIntyre, R. S., (2014). The integrative management of treatment-resistant depression: A comprehensive review and perspectives. *Psychotherapy and Psychosomatics, 83*, 70–88.

19. Chacón-Lee, T., & González-Mariño, G., (2010). Microalgae for "healthy" foods—Possibilities and challenges. *Comprehensive Reviews in Food Science and Food Safety, 9*, 655–675.
20. Chan, S. S., Khoo, K. S., Chew, K. W., Ling, T. C., & Show, P. L., (2022). Recent advances biodegradation and biosorption of organic compounds from wastewater: Microalgae-bacteria consortium: A review. *Bioresource Technology, 344*, 126159.
21. Chisti, Y., (2020). Microalgae biotechnology: A brief introduction. *Handbook of Microalgae-Based Processes and Products* (pp. 3–23). Elsevier.
22. Da Silva, V. B., Moreira, J. B., De Morais, M. G., & Costa, J. A. V., (2016). Microalgae as a new source of bioactive compounds in food supplements. *Current Opinion in Food Science, 7*, 73–77.
23. Das, M., & Keshri, J. P., (2017). Algal diversity in foot hills of Eastern Himalayas-II (*Cyanoprokaryota: Oscillatoriales*). *Phykos, 47*, 31–51.
24. Depraetere, O., Pierre, G., Noppe, W., Vandamme, D., Foubert, I., Michaud, P., & Muylaert, K., (2015). Influence of culture medium recycling on the performance of *Arthrospira platensis* cultures. *Algal Research, 10*, 48–54.
25. Dragone, G., Fernandes, B. D., Vicente, A. A., & Teixeira, J. A., (2010). *Third Generation Biofuels from Microalgae*, pp. 1355–1366.
26. Dwivedi, R. K., (20201). On some freshwater soft algae from Bhulla Taal Lake Lansdowne, District Pauri Garhwal, Uttarakhand. *J. Mountain Res., 16*, 341–348.
27. Farrelly, D. J., Everard, C. D., Fagan, C. C., & McDonnell, K. P., (2013). Carbon sequestration and the role of biological carbon mitigation: A review. *Renewable and Sustainable Energy Reviews, 21*, 712–727.
28. Fernando, I. S., Kim, M., Son, K. T., Jeong, Y., & Jeon, Y. J., (2016). Antioxidant activity of marine algal polyphenolic compounds: A mechanistic approach. *Journal of Medicinal Food, 19*, 615–628.
29. Gantar, M., & Svirčev, Z., (2008). Microalgae and cyanobacteria: Food for thought 1. *Journal of Phycology, 44*, 260–268.
30. Gao, K., (1998). Chinese studies on the edible blue-green alga, *Nostoc flagelliforme*: A review. *Journal of Applied Phycology, 10*, 37–49.
31. Grewe, C. B., & Pulz, O., (2012). The biotechnology of cyanobacteria. *Ecology of Cyanobacteria II* (pp. 707–739). Springer.
32. Grima, E. M., Belarbi, E. H., Fernández, F. A., Medina, A. R., & Chisti, Y., (2003). Recovery of microalgal biomass and metabolites: Process options and economics. *Biotechnology Advances, 20*, 491–515.
33. Guo, J. Y., Han, C. C., & Liu, Y. M., (2010). A contemporary treatment approach to both diabetes and depression by cordyceps sinensis, rich in vanadium. *Evidence-Based Complementary and Alternative Medicine, 7*, 387–389.
34. Gupta, P. L., Lee, S. M., & Choi, H. J., (2015). A mini review: Photobioreactors for large scale algal cultivation. *World Journal of Microbiology and Biotechnology, 31*, 1409–1417.
35. Hammond, P., (1992). Species inventory. *Global Biodiversity* (pp. 17–39). Springer.
36. Hansen, R. A., Gartlehner, G., Lohr, K. N., Gaynes, B. N., & Carey, T. S., (2005). Efficacy and safety of second-generation anti-depressants in the treatment of major depressive disorder. *Annals of Internal Medicine, 143*, 415–426.

37. Hu, Q., (2004). Industrial production of microalgal cell mass and secondary products-major industrial species. *Handbook of Microalgal Culture: Biotechnology and Applied Phycology* (p. 264).
38. Huchette, S. M. H., & Beveridge, M. C. M., (2005). Periphyton-based cage aquaculture. *Periphyton: Ecology, Exploitation, and Management*, 237–245.
39. Ismaiel, M. M. S., El-Ayouty, Y. M., & Piercey-Normore, M., (2016). Role of pH on antioxidants production by spirulina (*Arthrospira*) platensis. *Brazilian Journal of Microbiology, 47*, 298–304.
40. Jaiswal, K. K., Dutta, S., Banerjee, I., Pohrmen, C. B., & Kumar, V., (2021). Photosynthetic microalgae–based carbon sequestration and generation of biomass in biorefinery approach for renewable biofuels for a cleaner environment. *Biomass Conversion and Biorefinery*, 1–19.
41. Jiang, B., Xiong, Z., Yang, J., Wang, W., Wang, Y., Hu, Z. L., Wang, F., & Chen, J. G., (2012). Anti-depressant-like effects of ginsenoside Rg1 are due to activation of the BDNF signaling pathway and neurogenesis in the hippocampus. *British Journal of Pharmacology, 166*, 1872–1887.
42. Jindal, R., & Prajapat, P., (2005). Productivity and trophic status of Renuka Wetland (Distt. Sirmour, Himachal Pradesh). *Indian J. Ecol., 32*, 180–183.
43. Jindal, R., Thakur, R., Singh, U. B., & Ahluwalia, A., (2014). Phytoplankton dynamics and species diversity in a shallow eutrophic, natural mid-altitude lake in Himachal Pradesh (India): Role of physicochemical factors. *Chemistry and Ecology, 30*, 328–338.
44. Johansen, J. R., (2010). *Diatomsofaerial Habitats. The Diatoms: Applications for the Environmental and Earth Sciences*, 465.
45. Kaur, A., Singh, R., & Singh, R., (2018). Effect of altitudinal and seasonal factors on diatomological mapping of waterbodies: Implication in drowning investigations. *Arab Journal of Forensic Sciences & Forensic Medicine, 1*, 1013–1024.
46. Kim, N. H., Jeong, H. J., Lee, J. Y., Go, H., Ko, S. G., Hong, S. H., Kim, H. M., & Um, J. Y., (2008). The effect of hydrolyzed spirulina by malted barley on forced swimming test in ICR mice. *International Journal of Neuroscience, 118*, 1523–1533.
47. Koru, E., (2012). Earth food spirulina (*Arthrospira*): Production and quality standards. *Food Additive*, 191–202.
48. Kumar, R., Seth, M., & Suseela, M., (2015). *Xanthophyceae* and *Euglenophyceae* of Kangra District of Himachal Pradesh. *The Journal of Indian Botanical Society, 94*, 275–278.
49. Lafarga, T., Fernández-Sevilla, J. M., González-López, C., & Acién-Fernández, F. G., (2020). Spirulina for the food and functional food industries. *Food Research International, 137*, 109356.
50. Lafarga, T., Sánchez-Zurano, A., Villaró, S., Morillas-España, A., & Acién, G., (2021). Industrial production of spirulina as a protein source for bioactive peptide generation. *Trends in Food Science & Technology, 116*, 176–185.
51. Lembi, C. A., & Waaland, J. R., (1988). *Algae and Human Affairs*. Cambridge University Press.
52. Li, Y., Horsman, M., Wu, N., Lan, C. Q., & Dubois-Calero, N., (2008). Biofuels from microalgae. *Biotechnology Progress, 24,* 815–820.
53. Liu, F., Zhang, C., Wang, Y., & Chen, G., (2022). A review of the current and emerging detection methods of marine harmful microalgae. *Science of The Total Environment*, 152913.

54. Sanghvi, A. M., & Martin Lo, Y., (2010). Present and potential industrial applications of macro-and microalgae. *Recent Patents on Food, Nutrition & Agriculture, 2*, 187–194.
55. Mankotia, B., Sharma, P., Shekhar, J., Sood, G., & Thakur, R., (2011). *Upland Rice in Himachal Pradesh* (p. 107). Upland Rice in India.
56. Martinez-Goss, M. R., Arguelles, E. D., Sapin, A. B., & Almeda, R. A., (2021). Chemical composition and *in vitro* antioxidant and antibacterial properties of the edible cyanobacterium, nostoc commune vaucher. *Philippine Science Letters, 14,* 25–35.
57. Maso, M., & Garcés, E., (2006). Harmful microalgae blooms (HAB); problematic and conditions that induce them. *Marine Pollution Bulletin, 53*, 620–630.
58. Matufi, F., & Choopani, A., (2020). Spirulina, food of past, present and future. *Health Biotechnology and Biopharma, 3*, 1–20.
59. Moradi-Kor, N., Ghanbari, A., Rashidipour, H., Bandegi, A. R., Yousefi, B., Barati, M., Kokhaei, P., & Rashidy-Pour, A., (2020). Therapeutic effects of spirulina platensis against adolescent stress-induced oxidative stress, brain-derived neurotrophic factor alterations and morphological remodeling in the amygdala of adult female rats. *Journal of Experimental Pharmacology, 12*, 75.
60. Mulders, K. J., Lamers, P. P., Martens, D. E., & Wijffels, R. H., (2014). Phototrophic pigment production with microalgae: Biological constraints and opportunities. *Journal of Phycology, 50*, 229–242.
61. Munir, M., Qureshi, R., Ilyas, M., Munazir, M., & Leghari, M., (2016). Systematics of *Chroococcus* from Pakistan. *Pak. J. Bot., 48*, 255–262.
62. Nagaraj, S., (2019). *ALGAE (Biodiversity, Biology and Biotechnology)* (pp. 41–63). Center for Advanced Studies in Botany.
63. Oslan, S. N. H., Tan, J. S., Oslan, S. N., Matanjun, P., Mokhtar, R. A. M., Shapawi, R., & Huda, N., (2021). Haematococcus pluvialis as a potential source of astaxanthin with diverse applications in industrial sectors: Current research and future directions. *Molecules, 26*, 6470.
64. Panahi, Y., Badeli, R., Karami, G. R., Badeli, Z., & Sahebkar, A., (2015). A randomized controlled trial of 6-week *Chlorella vulgaris* supplementation in patients with major depressive disorder. *Complementary Therapies in Medicine, 23*, 598–602.
65. Panahi, Y., Darvishi, B., Jowzi, N., Beiraghdar, F., & Sahebkar, A., (2016). *Chlorella vulgaris*: A multifunctional dietary supplement with diverse medicinal properties. *Current Pharmaceutical Design, 22*, 164–173.
66. Parodi, A., Leip, A., De Boer, I., Slegers, P., Ziegler, F., Temme, E. H., Herrero, M., et al., (2018). The potential of future foods for sustainable and healthy diets. *Nature Sustainability, 1*, 782–789.
67. Pérez-Lloréns, J. L., (2020). Microalgae: From staple foodstuff to avant-garde cuisine. *International Journal of Gastronomy and Food Science, 21*, 100221.
68. Pirie, N., (1975). The spirulina algae. *Food Protein Sources, 4*, 33–39.
69. Ponnuswamy, I., Madhavan, S., & Shabudeen, S., (2013). Isolation and characterization of green microalgae for carbon sequestration, waste water treatment and bio-fuel production. *International Journal of Bio-Science and Bio-Technology, 5*, 17–26.
70. Posten, C., (2009). Design principles of photo-bioreactors for cultivation of microalgae. *Engineering in Life Sciences, 9*, 165–177.

71. Priyadarshani, I., & Rath, B., (2012). Commercial and industrial applications of micro algae: A review. *Journal of Algal Biomass Utilization, 3*, 89–100.
72. Pulz, O., (2001). Photobioreactors: Production systems for phototrophic microorganisms. *Applied Microbiology and Biotechnology, 57*, 287–293.
73. Ramaraj, R., Unpaprom, Y., & Dussadee, N., (2016). Cultivation of green microalga, *Chlorella vulgaris* for biogas purification. *International Journal of New Technology and Research, 2*, 263569.
74. Resende, D. I., Ferreira, M., Magalhães, C., Lobo, J. S., Sousa, E., & Almeida, I. F., (2021). Trends in the use of marine ingredients in anti-aging cosmetics. *Algal Research, 55*, 102273.
75. Rickard, L. N., Noblet, C. L., Duffy, K., & Christian, B. W., (2018). Cultivating benefit and risk: Aqua-culture representation and interpretation in New England. *Society & Natural Resources, 31*, 1358–1378.
76. Saeed, M. U., Hussain, N., Shahbaz, A., Hameed, T., Iqbal, H. M., & Bilal, M., (2021). Bioprospecting microalgae and cyanobacteria for biopharmaceutical applications. *Journal of Basic Microbiology*.
77. Sarial, A. K., (2019). Challenges and opportunities in crop diversification. *Himachal Journal of Agricultural Research, 45*, 1–14.
78. Sasaki, K., Othman, M. B., Demura, M., Watanabe, M., & Isoda, H., (2017). Modulation of neurogenesis through the promotion of energy production activity is behind the anti-depressant-like effect of colonial green alga, *Botryococcus braunii*. *Frontiers in Physiology, 8*, 900.
79. Shah, M. M. R., Liang, Y., Cheng, J. J., & Daroch, M., (2016). Astaxanthin-producing green microalga *Haematococcus pluvialis*: From single cell to high value commercial products. *Frontiers in Plant Science, 7*, 531.
80. Sharathchandra, K., & Sridhar, K. R., (2022). Dynamics of cyanobacteria in two thermal water bodies of the Western Ghats. *Expanding Horizon of Cyanobacterial Biology* (pp. 239–258). Elsevier.
81. Sharma, N., Sharma, P., & Sharma, N., (2020). A review on algal biodiversity of North Western Indian Himalayan Hillock–Himachal Pradesh and its potential as an attractive feedstock for 3rd generation biofuels. *ESSENCE Int. J. Env. Rehab. Conserv. XI (SP2)*, 135–144.
82. Sharoba, A. M., (2014). Nutritional value of spirulina and its use in the preparation of some complementary baby food formulas. *Journal of Food and Dairy Sciences, 5*, 517–538.
83. Shimizu, Y., (1996). Microalgal metabolites: A new perspective. *Annual Review of Microbiology, 50*, 431–466.
84. Sili, C., Torzillo, G., & Vonshak, A., (2012). *Arthrospira* (*Spirulina*). *Ecology of Cyanobacteria II* (pp. 677–705). Springer.
85. Singh, R., Upadhyay, A. K., Singh, D. V., Singh, J. S., & Singh, D. P., (2019). Photosynthetic performance, nutrient status and lipid yield of microalgae *Chlorella vulgaris* and *Chlorococcum humicola* under UV-B exposure. *Current Research in Biotechnology, 1*, 65–77.
86. Singh, U. B., & Ahluwalia, A., (2013). Microalgae: A promising tool for carbon sequestration. *Mitigation and Adaptation Strategies for Global Change, 18*, 73–95.

87. Singh, Y., Khattar, J., Singh, D., Rahi, P., & Gulati, A., (2014). Limnology and cyanobacterial diversity of high-altitude lakes of Lahaul-Spiti in Himachal Pradesh, India. *Journal of Biosciences, 39*, 643–657.
88. Soetantyo, G. I., & Sarto, M., (2019). The anti-depressant effect of *Chlorella vulgaris* on female Wistar rats (Rattus norvegicus Berkenhout, 1769) with chronic unpredictable mild stress treatment. *J. Trop. Biodivers. Biotechnol., 4*, 72–81.
89. Sommi, R. W., Crismon, M. L., & Bowden, C. L., (1987). Fluoxetine: A serotonin-specific, second-generation anti-depressant. *Pharmacotherapy: The Journal of Human Pharmacology and Drug Therapy, 7*, 1–14.
90. Cervantes-Llanos, M., Lagumersindez-Denis, N., Marín-Prida, J., Pavón-Fuentes, N., Falcon-Cama, V., Piniella-Matamoros, B., & Pentón-Rol, G., (2018). Beneficial effects of oral administration of C-phycocyanin and phycocyanobilin in rodent models of experimental autoimmune encephalomyelitis. *Life Sciences, 194*, 130–138.
91. Subermaniam, K., Teoh, S. L., Yow, Y. Y., Tang, Y. Q., Lim, L. W., & Wong, K. H., (2021). Marine algae as emerging therapeutic alternatives for depression: A review. *Iranian Journal of Basic Medical Sciences, 24*, 997.
92. Suganya, T., Varman, M., Masjuki, H., & Renganathan, S., (2016). Macroalgae and microalgae as a potential source for commercial applications along with biofuels production: A biorefinery approach. *Renewable and Sustainable Energy Reviews, 55*, 909–941.
93. Suresh, D., Madhu, M., Ch, S., & Shankaraiah, P., (2014). Anti-depressant activity of spirulina platensis in experimentally induced depression in mice. *International Journal of Research and Development in Pharmacy & Life Sciences, 3,* 1026–1035.
94. Talbott, S., Hantla, D., Capelli, B., Ding, L., Li, Y., & Artaria, C., (2019). Astaxanthin supplementation reduces depression and fatigue in healthy subjects. *EC Nutrition, 14*, 239–246.
95. Taylor, W. D., (2014). Depression in the elderly. *New England Journal of Medicine, 371*, 1228–1236.
96. Thakur, R., Jindal, R., Singh, U. B., & Ahluwalia, A., (2013). Plankton diversity and water quality assessment of three freshwater lakes of Mandi (Himachal Pradesh, India) with special reference to planktonic indicators. *Environmental Monitoring and Assessment, 185*, 8355–8373.
97. Tirichine, L., & Bowler, C., (2011). Decoding algal genomes: Tracing back the history of photosynthetic life on Earth. *The Plant Journal, 66*, 45–57.
98. Volkman, J. K., (2014). Acyclic isoprenoid biomarkers and evolution of biosynthetic pathways in green microalgae of the genus *Botryococcus*. *Organic Geochemistry, 75*, 36–47.
99. Zeng, X., Zhang, Y., Kwong, J. S., Zhang, C., Li, S., Sun, F., Niu, Y., & Du, L., (2015). The methodological quality assessment tools for preclinical and clinical studies, systematic review and meta-analysis, and clinical practice guideline: A systematic review. *Journal of Evidence-Based Medicine, 8*, 2–10.

## CHAPTER 9

# Antidepression Agents in Himalayan Spices

AHMED OLATUNDE,[1] HABIBU TIJJANI,[2] and
BASIRU OLAITAN AJIBOYE[3]

[1]*Department of Medical Biochemistry, Abubakar Tafawa Balewa University, Bauchi, Nigeria*

[2]*Department of Environmental Health Science, National Open University of Nigeria, Abuja, Nigeria*

[3]*Phytomedicine and Molecular Toxicology Research Laboratory, Department of Biochemistry, Federal University Oye-Ekiti, Nigeria*

## ABSTRACT

Depression, a neurodegenerative disease, is characterized by loss of interest, reduced concentration, decreased energy, and other neurodegenerative-related symptoms. The pathogenesis of the disease involves alterations in brain monoamine levels, impaired hypothalamic-pituitary-adrenal axis, neurodegeneration, nitrative and oxidative stress, immune-inflammatory processes, and inhibition of neurogenesis. Various antidepressant agents, such as serotonin-norepinephrine reuptake inhibitors, tricyclic antidepressant drugs, selective serotonin receptor inhibitors, and monoamine oxidase inhibitors are used for managing depression and its symptoms. However, these synthetic antidepressive agents have unwanted side effects and do not fully attenuate the pathological characteristics of depression. Therefore, the search for alternative treatment strategies has shifted towards the use of medicinal plants. The Himalayan region is rich in a wide range

Himalayan Medicinal Plants for the Treatment of Depression: A Source of Rich Antidepressant Agents. Abdur Rauf, Prabhakar Semwal, & Hafiz Ansar Rasul Suleria (Eds.)

of medicinal plants. Spices and their secondary metabolites found in this region have been documented to be effective in managing depression and associated neurodegenerative disorders. Antidepressant compounds from Himalayan spices can be classified into carotenoids, flavonoids, phenolic acids, polyphenols, and others. Crocin, ferulic acid, kaempferol, crocetin, gallic acid, quercetin, and rutin are examples of these compounds with antidepressant activity. Therefore, this work highlights the potential of using bioactive compounds from Himalayan spices in the management of depression and related diseases.

## 9.1 INTRODUCTION

Depression is a disorder that typically includes symptoms such as anxiety, loss of interest, loss of appetite, sleep disturbances, suicidal thoughts, and lack of energy [1]. Stress is a major contributor to depression, which is prevalent worldwide. It manifests as the body's response to various stimuli, whether physical, mental, or emotional [2]. Depression is triggered by stress and challenging conditions that are difficult to alleviate, leading to impaired functioning and suffering in daily life activities. In severe cases, it can even result in death. Additionally, the World Health Organization (WHO) has stated that the American Psychiatric Association's Diagnostic and Statistical Manual (DSM-V) and the International Statistical Classification of Diseases and Related Health Problems (ICD-10) have documented different types of depression, categorized based on severity and the number of symptoms. These include major to severe depression, mild to moderate depression, and bipolar affective disorder [1, 3, 4]. From another dimension, the National Institute of Mental Health has documented that depression can manifest in various ways: seasonal affective disorder, perinatal depression, disruptive mood impairment, and premenstrual dysphoric disorder [5–8]. Depression is a neurodegenerative disorder and is one of the leading causes of distress worldwide. The World Health Organization (WHO) has stated that over 300 million individuals experience depressive symptoms, accounting for approximately 4.4% of the global population [9]. Furthermore, between 2005 and 2015, there was an 18% increase in reported cases of depression, with women being affected 50% more than men globally. As a result, depression poses a significant disease burden for females in low-, middle-, and high-income countries [1, 9].

Medicinal plants and their products have been proven to be effective and have a wide range of applications in the treatment of depression. They can be used as a supplementary therapy or even as an alternative treatment strategy for depression [10]. The Himalayan region is known for its rich diversity of natural resources and their products. This region greatly contributes to the livelihood of local communities and villages. Furthermore, numerous studies have shown that Himalayan plants contain powerful bioactive compounds, minerals, and vitamins. The phytochemicals present in these plants have a significant role in traditional herbal systems [11]. Spices are one of the types of plants found in this region. Therefore, this chapter aims to highlight the use of biologically active compounds found in spices from the Himalayan region for the treatment of depression and related neurodegenerative diseases.

## 9.2 NEUROCHEMISTRY OF DEPRESSION

The monoamine theory of depressive behavior has been a main target of research in the area of pharmacotherapy and pathophysiology for over 25 years, and the majority of antidepressant management is based on this proposal [12]. The theory suggests that suppression in the levels of brain NE, serotonin, and/or dopamine is the major basis of depression [13]. Dopamine, serotonin (5-hydroxytryptamine (5-HT)), and norepinephrine are the main neurotransmitters in the brain. Serotonin, in particular, primarily affects moods and depression. Serotonin binding sites are mainly found in the brain's raphe nucleus and other locations such as the limbic and striatum system, the frontal cortex, affecting the hippocampus and hypothalamus [14]. This neurotransmitter mediates appetite, violence, hunger, anxiety, perception, impulsiveness, fearfulness, and other functions. It also stimulates memory retrieval and formation. On the other hand, dopamine mainly acts in reward-motivated patterns, motor control, and the release of various second messengers, among other activities [15]. Serotonin receptors regulate the formation of several second messengers and neurotransmitters. They also affect neurological and biological processes, including mood, learning, anxiety, appetite, sleep, cognition, and nausea [1, 15]. Serotonin is divided into seven classes: serotonin-1, -2, -3, -4, -5, -6, and -7. Furthermore, these classes are further subdivided into serotonin-1F, -1E, -1D, -1B, -1A, -2C, -2B, -2A, -3, -4, -5B, -5A, -6, and -7 [16]. Therefore, they have been utilized as targets for different therapeutic agents.

Research has documented that individuals with major depression exhibit alterations in serotonin-1A receptors as well as a decrease in its density at the postsynaptic terminal [17, 18]. Genetic studies have documented that individuals with high activity and density of serotonin-1A autoreceptors are more susceptible to mood disorders and respond inadequately to treatment with antidepressants [19]. Based on this, it has been suggested that serotonin-1A antagonists may exhibit antidepressant action. Activation of serotonin-1B heteroreceptors enhances the antidepressant property, and serotonin-1B receptor knockout mice were reported to exhibit significantly disordered behavior and have an elevated preference for alcohol [20, 21]. Thus, the activation of the serotonin-1B receptor may play an essential function in antidepressant action. Unlike the receptors of serotonin-1B and serotonin-A, the clinical importance of other subclasses of serotonin receptors is still elusive [1].

Other receptors, such as adrenergic or noradrenergic receptors (consisting of alpha and beta), dopamine receptors, and others, also play an essential role in the antidepressant activity of therapeutic agents. Additionally, the inhibition of enzymes, including monoamine oxidases, which are enzymes that induce oxidative deamination of amines such as NE, serotonin, and dopamine in neurons, by different chemical agents and bioactive compounds from plants, inhibits the degradation process of neurotransmitters. This elevation in their levels in the brain causes an antidepressant action [22].

## 9.3 ROLE OF PLANTS AND THEIR COMPOUNDS IN TREATING DEPRESSION

The application of medicinal plants is still vital in basic healthcare, and several current drugs are designed from plants [1, 23]. Medicinal plants and traditional remedies have gained attention in the management of depression and related disorders. Approximately three-quarters of the global population rely on medicinal plants for their basic healthcare [24]. The therapeutic importance of plants with medicinal properties is evident in the percentage of medical prescriptions, of which 25% are obtained from plants. It has been documented that 157 out of 520 therapeutic agents are of plant origin [25, 26]. Supporting this study, it has been shown that more than 75% of these compounds are utilized in treating infectious diseases [27]. Several studies have revealed that many secondary metabolites, such as alkaloids, saponins, polyphenols, flavonoids, essential oils, and fatty acids, exhibit antidepressant and anxiolytic properties [28].

## 9.4 ANTIDEPRESSANT AGENTS FROM PLANT SPICES FOUND IN THE HIMALAYAN REGION

### *9.4.1 FLAVONOIDS*

Kaempferol, a flavonoid, has a broad range of therapeutic actions, such as antidepressant effects. Kaempferol obtained from the tepals of saffron (*Crocus sativus*), a spice found in the Himalayan region, has been reported to be responsible for its activity against depression in mice and rat models [29]. In this study, doses of 200, 100, and 50 mg/kg body weight of the compound decreased depression symptoms, which was comparable to a reference drug, fluoxetine [30]. In a mouse model with chronic social defeat stress, kaempferol remarkably increased consumption of sucrose, body weights, mobility time of tail suspension test (TST), and social interaction time, thus exhibiting antidepressant actions. The authors stated that this mechanism of action of the compound was achieved by elevating the actions of superoxide dismutase (SOD), catalase, and glutathione transferase, and suppressing the concentrations of inflammatory regulators and activation of microglia, including tumor necrosis factor-α and interleukin-1β, as well as stimulating and promoting the action of the prefrontal cortex Akt/β-catenin pathway [31]. Additionally, it was stated that kaempferol attenuated symptoms of depression by modulating ESR1, monoamine oxidase-B, and monoamine oxidase-A, which are vital target genes of different mental disorders and specifically vital for depression [32].

Quercetin is a flavonoid that has been reported to have antidepressant effects in animal models. In various tests such as the forced swim test (FST), sucrose consumption test, and TST, this flavonoid has shown improvement in depressive symptoms in rats subjected to unpredictable chronic mild stress and lipopolysaccharide-induced depression [33–35]. The antidepressant effects are attributed to its anti-inflammatory and antioxidant properties, which include the suppression of hippocampal nitric oxide and TBARS levels, reduction of malondialdehyde levels, elevation of serotonin levels, decrease in glutathione and total thiol levels, and modulation of SOD, catalase, and glutamate activities. Quercetin also reduces the release of cyclooxygenase-2, tumor necrosis factor-α, interleukin-1β, nuclear factor-kappa B, and interleukin-6, while decreasing corticosterone levels in serum [34–36]. Furthermore, quercetin has been found to restore the expression of copine 6, brain-derived neurotrophic factor, p-TrkB, and stimulating receptors expressed over myeloid cells 1 and 2 in the hippocampus and prefrontal cortex, thereby improving depressive behaviors, memory, and learning impairments in lipopolysaccharide-treated rats [33]. In an animal model of depression induced by olfactory bulbectomy, quercetin treatment was shown to alleviate elevated levels of lipid hydroperoxide and exhibit antidepressant effects by inhibiting the activity of nitric oxide synthase and the biosynthesis of N-methyl-D-aspartate receptors and nitric oxide [37]. These mechanisms highlight the antidepressant properties of quercetin, a flavonoid found in *C. sativus,* which is an example of spices found in the Himalayan region.

Quercetin-3-rutinoside, also known as rutin, consists of the flavanol quercetin and rutinose [38]. In a rat model of chronic unpredictable stress-induced depression, treatment with quercetin-3-rutinoside was found to improve the morphology and cell number of the hippocampus, similar to that of normal rats. This treatment conferred protection against the loss of hippocampal neurons [39]. Similarly, the compound exhibited antidepressant effects through its neuroprotective activity on the hippocampus. It inhibited NMDA receptor molecules via its NR2A and NR2B subunits, regulated the percentage of dark neurons, and influenced the diameter of the CA3 region in the hippocampus of mice subjected to maternal separation stress [40]. Another study demonstrated that quercetin-3-rutinoside displayed antidepressant action by modulating the noradrenergic and serotonergic and/or dopaminergic systems in animals treated with α-methyl-p-tyrosine and p-chlorophenylalanine methyl ester [41] (Figure 9.1).

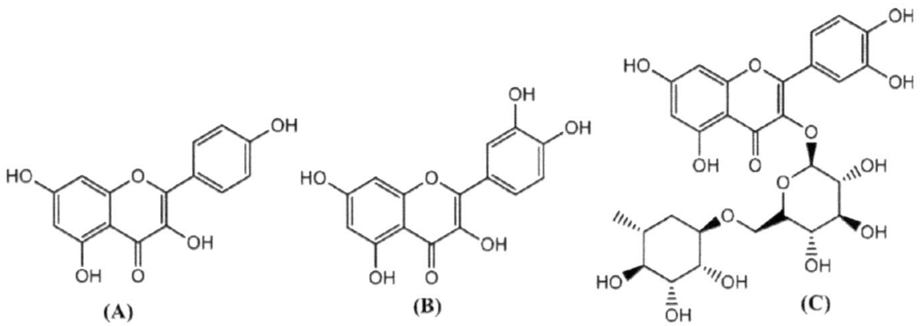


**FIGURE 9.1** Structure of (A) kaempferol; (B) quercetin; and (C) rutin.

### 9.4.2 PHENOLIC ACID

Ferulic acid, found in *Ligusticum chuanxiong* Hort, has been shown to possess antidepressant action. However, the main pharmacological mechanistic actions involved are still elusive. The phenolic acid has been reported to enhance oxidative stress and corticosterone-induced alterations in behavior [42]. In another study, ferulic acid remarkably attenuated depression in chronic unpredictable mild stress (CUMS) mice. The compound was proposed to exert its actions against depression through an increase in the mRNA level of brain-derived neurotrophic factor (BDNF) and elevation of the concentration of synapsin I and postsynaptic cortex in the prefrontal cortex and hippocampus [43].

Gallic acid has been shown to be an effective compound in managing depression. In one study, doses of 50–150 mg/kg of phenolic acid were found to ameliorate hippocampal cellular loss induced by trimethyltin chloride, thereby reversing depression in experimental rats [44]. Another study found that doses of 10–20 mg/kg of gallic acid exhibited antidepressant action in mice exposed to unpredictable mild stress, by inhibiting monoamine oxidase-A activity and reducing plasma concentrations of corticosterone and nitrite [45]. Furthermore, doses of 25–50 mg/kg of gallic acid were found to alleviate symptoms of depression and reduce oxidative stress in post-stroke depressive balb/c mice [46]. In a recent study, gallic acid (50–100 mg/kg) demonstrated protective activity on impaired memory and depression in sodium arsenite-treated male rats, which was attributed to its antioxidant activity [47]. In conclusion, gallic acid exerts its antidepressant action by enhancing antioxidant enzyme activities, serotonin, and catecholamine levels in the central nervous system (CNS). It also modulates dopaminergic, alpha adrenergic, and serotonergic receptors [48, 49] (Figure 9.2).

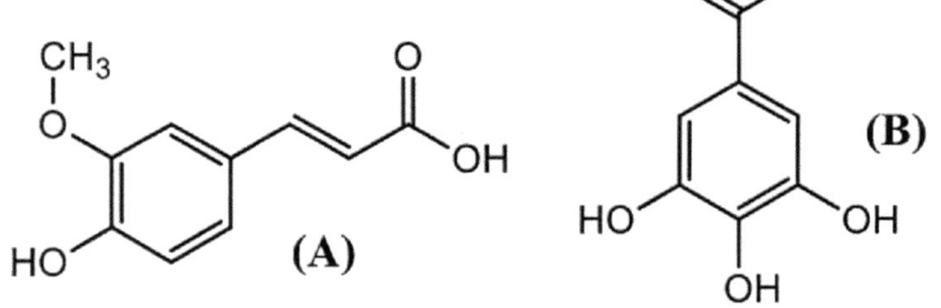


**FIGURE 9.2** Structure of (A) ferulic acid; and (B) gallic acid.

### 9.4.3 POLYPHENOLS

Polyphenols present in Himalayan spices have been documented to exhibit antidepressant effects. It has been demonstrated that Melissa officinalis shows anxiolytic effects due to the presence of ursolic acid, rosmarinic acid, pentacyclic triterpenes, and oleanolic acid. In the same study, rosmarinic acid was found to have stronger inhibitory effects against γ-aminobutyric acid (GABA) transaminase compared to other compounds in the plant [50]. This plant can be found in Western and Northern Asia [51].

Curcumin has been evidenced for its role against depression in both animal and clinical studies. However, the human trials of the compound against depression are still controversial [52, 53]. Curcumin has been shown to elevate antioxidant enzyme actions and alleviate the symptoms of depression in an unpredictable chronic mild stress animal model. The compound also modulates cholinergic action, leading to subsequent memory function improvement [54]. In consistent findings, chronic curcumin treatment remarkably attenuated symptoms of depression by preventing neuronal apoptosis induced by interleukin-1β within ventromedial prefrontal cortex neurons in unpredictable chronic mild stress-induced rats [55]. Additionally, chronic curcumin treatment elevated brain-derived neurotrophic factor levels in the amygdala and hippocampi through the elevation of ERK phosphorylation, thus alleviating depression in animal models [56–58]. In a similar study, curcumin prevented reduced synaptophysin protein and brain-derived neurotrophic factor expression, and attenuated PSD-95 expression, which was linked to its antidepressant activity in CUMS-induced animals [59]. The compound also improved the concentrations of central monoaminergic neurotransmitters such as 3,4-dihydroxyphenylacetate, noradrenaline, 5-hydroxyindoleacetate, and serotonin in the hippocampal area, while enhancing concentrations of noradrenaline, 5-hydroxyindoleacetate, and dopamine in the frontal cortex of olfactory bulbectomy rat models subjected to the FST [60–62]. Curcumin has been reported to improve postsynaptic electrical reactivity and cell viability in neural circuits by regulating the expression of inflammatory proteins and brain-derived neurotrophic factors [63]. In animal models of depression induced by olfactory bulbectomy, curcumin has been proposed to decrease the levels of malondialdehyde and nitrite, elevate the level of corticosterone, and improve the decreased levels of catalase and reduced glutathione. Additionally, curcumin has been found to elevate inflammatory proteins such as tumor necrosis factor-α, apoptotic factors like caspase-3, and improve the expression of BDNFs [64]. Curcumin exhibits antidepressant actions by inhibiting the kynurenine cascade and NLRP3 inflammasome. The compound reduces the expression of proinflammatory proteins (interleukin-1β, tumor necrosis factor-α, and interleukin-6), improves the levels of inducible nitric oxide synthase and cyclooxygenase 2, reduces nuclear factor-γ B and microglial activation, indolamine-2, 3-dioxygenase, and the P2X7R/NLRP3 inflammasome axis induced by stress. Furthermore, curcumin attenuates the elevated kynurenine/tryptophan ratio [65–67]. Treatment with curcumin has been shown to alleviate symptoms of depression and improve brain metabolism by inducing potent deactivation of the left primary auditory cortex and stimulation of the amygdalohippocampal cortex in animal models treated with chronic unpredictable stress [68]. Another study suggests that the antidepressant mechanism of action of curcumin may involve its ability to inhibit the release of glutamate from the prefrontal cortex nerve terminals of rats by reducing presynaptic $Ca_v2.1$ (P/Q-type) and $Ca_v2.2$ (N-type) channels, rather than by blocking $Na^+/Ca^{2+}$ exchange or release of intracellular $Ca^{2+}$ [69] (Figure 9.3).

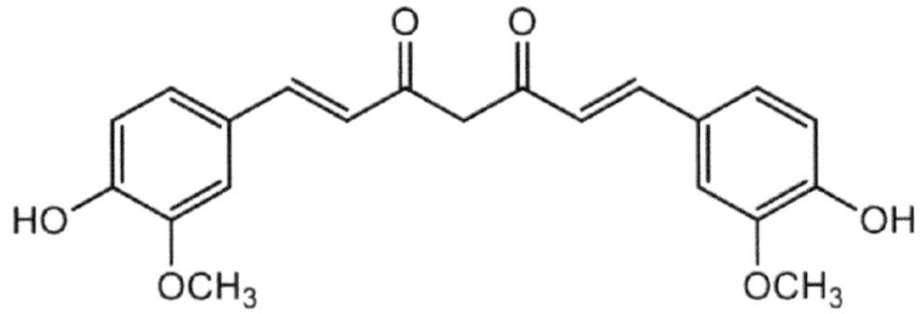


**FIGURE 9.3** Structure of curcumin.

### *9.4.4 CAROTENOIDS*

Crocin, found in the *C. sativus* plant, has been reported to account for the antidepressant activity of the plant's aqueous extract [70, 71]. Prior to the aforementioned study, Rios et al. [72] indicated that safranal and crocin, derived from the saffron plant, exhibited antidepressant effects in an *in vivo* study. They demonstrated that these two compounds increased stereotypic action and climbing time, and concluded that safranal and crocin from *C. sativus* were responsible for the antidepressant activity [72, 73]. In a similar study, safranal and crocin from

*C. sativus* alleviated symptoms of depressive behavior, similar to the effects of two antidepressant agents, imipramine and fluoxetine [74]. Safranal also demonstrated memory-enhancing and anticonvulsant properties through its action on the central nervous system [75]. Another study stated that crocin and safranal from *C. sativus* were responsible for its antidepressant effects in a mouse model [76]. Consistent with this finding, it was shown that the antioxidant activity of crocin influenced serotonergic receptors, thereby improving symptoms of obsessive-compulsive disorder, a type of depression [77]. Intraperitoneal injection of crocin (50 mg/kg) exerted anxiolytic effects in experimental rats, with comparable results to diazepam [78]. Furthermore, investigation on anxiety disorder indicated that safranal at doses of 0.15 and 0.35 mg/kg body weight exerted antianxiety activity [78,79]. In clinical trials, crocin (30 mg/kg) in a double-blind, placebo-controlled pilot human trial exerted antidepressant and antianxiety activities [80]. In conclusion, it has been documented that the antidepressant action of compounds from saffron, a spice found in the Himalayan region, is mainly via serotonin reuptake inhibition, abrogation of reuptake of NE and dopamine, and others [76, 81, 82].

In a study, crocin (100 mg/kg) was orally administered and remarkably attenuated the immobility time in the FST without influencing the initial locomotion of mice [83]. The compound also exhibited antidepressant action in a chronic corticosterone-induced depression animal model, as observed in the open field test (OFT), TST, and FST [84]. The underlying mechanistic action involved in these tests is linked to the improved expression of hippocampal cAMP response element binding protein, p-cAMP response element binding protein, brain-derived neurotrophic factor protein, and vascular growth factor in rats. In the study, crocin demonstrated an antidepressant effect by protecting the dopaminergic projection neurons located in the ventral tegmental region in mice with sub-acute Parkinson's disease induced by 1-methyl-4-phenyl-1,2,3,6-tetrahydropyridine. This action was related to the stimulation of mammalian target of rapamycin, resulting in the enhancement of neural synaptic plasticity in the medial prefrontal cortex [85, 86]. It was also reported that crocin decreased the expression levels of SOD family and proinflammatory cytokines such as interleukin-1β, thereby reducing oxidative stress and neuroinflammation in the hippocampi, alleviating symptoms of depression. Crocin decreased nicotinamide accumulation, enhanced the synthesis of nicotinamide adenine dinucleotide (oxidized), and activated SIRT3 deacetylase activity, as well as increased antioxidant enzymes (SOD and glutathione reductase) activities. Additionally, the compound abrogated the levels of markers for oxidative damage, including malondialdehyde and reactive oxygen species. It also alleviated impaired mitochondrial activity and restored ATP synthesis level [84]. The administration of crocin was shown to decrease proinflammatory cytokines and alleviate depression in the hippocampus of a chronic obstructive pulmonary disease animal model. Moreover, it reduced peribronchial inflammatory cytokine infiltration, abrogated inflammatory cell quantity, and decreased the level of proinflammatory cytokines in lung tissue and bronchoalveolar lavage fluid via the modulation of the PI3K/Akt-regulated inflammatory cascade [87]. Consistently, the compound exerted activity against symptoms of depression induced by lipopolysaccharide through antioxidant and anti-inflammatory actions. In BV-2 cell lines, crocin abrogated nitric oxide synthesis and the formation of tumor necrosis factor-alpha and interleukin-1β, as well as reactive oxygen species. It also reduced the expression of nuclear factor-γ B p65, CD16/32, and inducible nitric oxide synthase, while increasing the mRNA level of CD206. In the hippocampus of mice induced by lipopolysaccharide, the expression of ASC, caspase-1, and NLRP3 was neutralized, resulting in a decrease in the concentration of interleukin-18, interleukin-1β, and tumor necrosis factor-β [88]. In another study involving prenatal stress in mice, the compound improved the expression of defective hippocampal synaptic plasticity-linked proteins by regulating the growth hormone secretagogue receptor-PI3K cascade, leading to a prolonged and fast-onset antidepressant activity [89].

Crocetin is a carotenoid found in the C. sativus plant. The compound has been documented for its antidepressant action. In view of this, the compound (50, 25, and 12.5 mg/kg) reduced the immobility time in TST and FST without affecting the initial coordination and locomotor action of experimental mice. Additionally, crocetin significantly attenuated the number of crossings and immobility time in chronic restraint stress rats by improving the levels of reduced glutathione and malondialdehyde, as well as the activities of antioxidant enzymes [83, 90] (Figure 9.4).

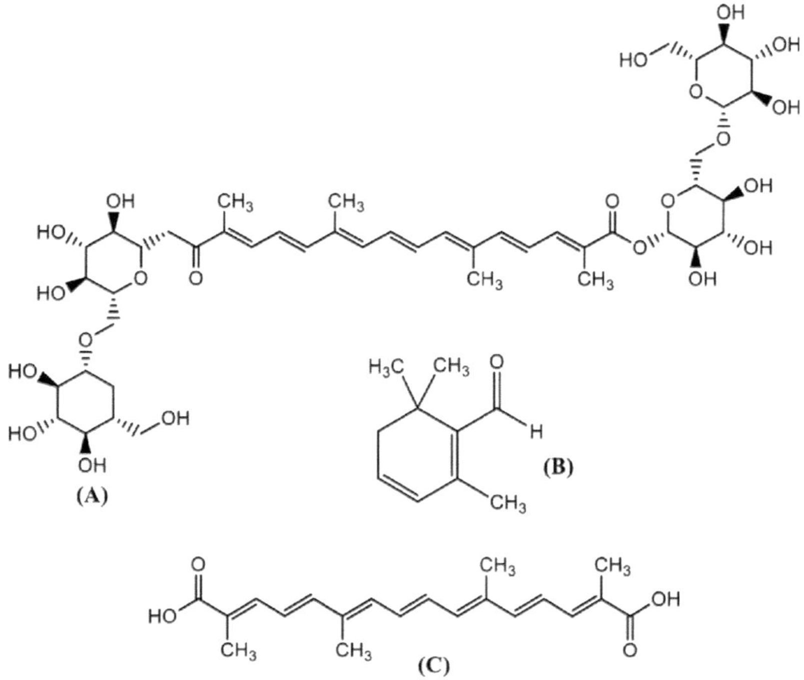

FIGURE 9.4 Structure of (A) crocin; (B) safranal; and (C) crocetin.

### 9.4.5 OTHER COMPOUNDS

Tetramethylpyrazine, isolated from *L. chuanxiong* Hort and found in the Sichuan region of China, has been reported to exhibit antidepressant effects. The compound was found to inhibit Toll-like receptor 4 and other related proteins in the hippocampus and prefrontal cortex of CUMS mice. Additionally, it was observed that the compound increased levels of NE and serotonin in both the brain and serum. Overall, the authors concluded that tetramethylpyrazine has the potential to act as an antidepressant through the inhibition of Toll-like receptor 4-nuclear factor-κB-nucleotide binding proteins in the mouse brain [91].

Another compound called ligustilide, found in the aforementioned plant, was also reported to display antidepressant-like activity. This was achieved by elevating the concentrations of allopregnanolone and progesterone in the prefrontal cortex and hippocampus of CUMS mice [92]. Additionally, ligustilide was shown to elevate the mRNA levels of hippocampal tryptophan and serotonin, resulting in its antidepressant-like action in CUMS rats [93].

## 9.5 CONCLUSION

Himalayan regions constitute a wide range of medicinal plants, including spices. These plants possess medicinal values and exert antidepressant actions through various mechanisms. These mechanisms include the inhibition of monoamine oxidase expression, improvement of monoamines and neurotransmitters (such as 3,4-dihydroxyphenylacetate, dopamine, 5-hydroxyindoleacetate, and serotonin) in the frontal cortex and hippocampus, inhibition of neuroinflammation by reducing the level of pro-inflammatory proteins (mostly interleukin-6, tumor necrosis factor, and interleukin-1), modulation of hippocampal brain-derived neurotrophic factor expression by stimulating its intracellular targets, suppression of oxidative biomarker synthesis, and enhancement of antioxidant enzyme activities. These mechanisms are exhibited by compounds found in these plants, including quercetin, crocin, gallic acid, kaempferol, rutin, among others. Therefore, these biologically active compounds present in Himalayan spices could serve as adjuvants or be used as a basis for synthesizing therapeutic agents for the treatment of depression and related neurodegenerative disorders.

## KEYWORDS


- **crocins**
- **hippocampus**
- **kaempferol**
- **monoamine oxidases**
- **neurodegenerative disorders**
- **neuroinflammation**
- **neurotransmitters**
- **serotonin receptors**

## REFERENCES


1. Martins, J., & Brijesh, S., (2018). Phytochemistry and pharmacology of anti-depressant medicinal plants: A review. *Biomedicine & Pharmacotherapy, 104*(2018), 343–365.
2. Jones, F., Bright, J., & Clow, A., (2001). *Stress: Myth, Theory and Research.* Pearson Education, Upper Saddle River, NJ: Prentice Hall.
3. World Health Organization, (2016). *International Statistical Classification of Diseases and Related Health Problems, 10th Revision (ICD-10). Mental and Behavioral Disorders (F00-F99).* Chapter V. http://apps.who.int/classifications/icd10/browse/2016/en#/F10-F19 (accessed on 22 January 2023).
4. American Psychiatric Association, (2013). *Diagnostic and Statistical Manual of Mental Disorders (DSM-V)* (5th edn., pp. 155–184). American Psychiatric Publishing. http://displus.sk/DSM/subory/dsm5.pdf (accessed on 22 January 2023).
5. O'Hara, M. W., & Swain, A. M., (1996). Rates and risk of postpartum depression: A meta-analysis. *International Review of Psychiatry, 8*, 37–54.
6. Dougherty, L. R., Smith, V. C., Bufferd, S. J., Kessel, E. M., Carlson, G. A., & Klein, D. N., (2016). Disruptive mood dysregulation disorder at the age of six years and clinical and functional outcomes six years later. *Psychological Medicine, 46*(5), 1103–1114.
7. Rapkin, A. J., & Lewis, E. I., (2013). Treatment of premenstrual dysphoric disorder. Women's health. *SAGE, 9*(6), 537–556.
8. National Institute of Mental Health, Mental Health, and Education, (2017). https://www.nimh.nih.gov/health/topics/depression/index.shtml (accessed on 22 January 2023).
9. World Health Organization, Depression Fact Sheet, (2017). http://www.who.int/mediacentre/factsheets/fs369/en/ (accessed on 22 January 2023).
10. Agha-Hosseini, M., Kashani, L., Aleyaseen, A., Ghoreishi, A., Rahmanpour, H., Zarrinara, A. R., et al., (2008). *Crocus sativus* L. (saffron) in the treatment of premenstrual syndrome: A double-blind, randomized and placebo-controlled trial. *British Journal of Obstetrics and Gynaecology, 115*(4), 515–519. doi: 10.1111/j.1471- 0528.2007.01652.x.
11. Semwal, P., Painuli, S., Jamloki, A., Rauf, A., Rahman, M. D., Olatunde, A., Hemeg, H. A., et al., (2022). Himalayan wild fruits as a strong source of nutraceuticals, therapeutics, food, and nutrition security. *Food Reviews International.* doi: 10.1080/87559129.2022.2121407.
12. Christopher, W., (1994). The serotonin theory of depression. Jeff. J. Psychiatry, 12. Article 4.
13. Schildkraut, J. J., (1965). The catecholamine hypothesis of affective disorders: A review of supporting evidence. *American Journal of Psychiatry, 122*(1965), 609–622.
14. Meneses, A., (1999). 5-HT system and cognition. *Neuroscience & Biobehavioral Review, 23*, 1111–1125.
15. Young, S. N., & Leyton, M., (2002). The role of serotonin in human mood and social interaction. Insight from altered tryptophan levels. *Pharmacology Biochemistry Behavior, 71*, 857–865.

16. Stahl, S. M., (1998). Mechanism of action of serotonin selective reuptake inhibitors serotonin receptors and pathways mediate therapeutic effects and side effects. *Journal of Affective Disorders, 51*, 215–235.
17. Boldrini, M., Underwood, M. D., Mann, J. J., & Arango, V., (2008). Serotonin-1A autoreceptor binding in the dorsal raphe nucleus of depressed suicides. *Journal of Psychiatric Research, 42*, 433–442.
18. Bhagwagar, Z., Rabiner, E. A., Sargent, P. A., Grasby, P. M., & Cowen, P. J., (2004). Persistent reduction in brain serotonin-1A receptor binding in recovered depressed men measured by positron emission tomography with [11C]WAY-100635. *Molecular Psychiatry, 9*, 386–392.
19. Lemonde, S., Turecki, G., Bakish, D., Hrdina, P. D., & Bown, C. D., & Sequeira, A., (2003). Impaired repression at a 5-hydroxytryptamine 1A receptor gene polymorphism associated with major depression and suicide. *Journal of Neuroscience, 23*, 8788–8799.
20. Chenu, F., David, D. J., Leroux-Nicollet, I., LeMaitre, E., Gardier, A. M., & Bourin, M., (2008). Serotonin 1B heteroreceptor activation induces an antidepressant-like effect in mice with an alteration of the serotonergic system. *Journal of Psychiatry & Neuroscience, 33*, 541–550.
21. Chilmonczyk, Z., Bojarski, A. J., Pilc, A., & Sylte, I., (2015). Functional selectivity and antidepressant activity of serotonin 1A receptor ligands. *International Journal of Molecular Science, 16*, 18474–18506.
22. Kalgutkar, A. S., Dalvie, D. K., Castagnoli, N., & Taylor, T. J., (2001). Interactions of nitrogen containing xenobiotics with monoamine oxidase (MAO) isozymes A and B: SAR studies on MAO substrates and inhibitors. *Chemical Research in Toxicology, 14*, 1139–1162.
23. UNESCO, (1996). *Culture and Health, Orientation Texts-World Decade for Cultural Development 1988–1997* (p. 129). Document CLT/DEC/PRO -1996, Paris, France.
24. World Health Organization, WHO, (2013). *Traditional Medicine Strategy, 54*, 2014–2023.
25. Duke, J. A., Janick, J., & Simon, J. E., (1993). *Medicinal Plants and the Pharmaceutical Industry* (pp. 664–669). Wiley, New Crops. New York.
26. Cragg, G. M., Newman, D. J., & Snader, K. M., (1997). Natural products in drug discovery and development. *Journal of Natural Products, 60*, 52–60.
27. Newman, D. J., Cragg, G. M., & Snader, K. M., (2003). Natural products as sources of new drugs over the period 1981–2002. *J. Nat. Prod., 66*, 1022–1037.
28. Fajemiroye, J. O., DaSilva, D. M., DeOliveira, D. R., & Costa, I. A., (2016). Treatment of anxiety and depression: Medicinal plants in retrospect. *Fundam. Clin. Pharmacol., 30*, 198–215.
29. Hosseinzadeh, H., Motamedshariaty, V., & Hadizadeh, F., (2007). Antidepressant effect of kaempferol, a constituent of saffron (*Crocus sativus*) petal, mice, and rats. *Pharmacology Online, 2*, 367–370.
30. Melnyk, J. P., Wang, S., & Marcone, M. F., (2010). Chemical and biological properties of the world's most expensive spice: Saffron. *Food Res. Int., 43*, 1981–1989.
31. Gao, W., Wang, W., Peng, Y., et al., (2019). Antidepressive effects of kaempferol mediated by reduction of oxidative stress, proinflammatory cytokines and

up-regulation of AKT/β-catenin cascade. *Metab. Brain Dis., 34*(2), 485–494. doi: 10.1007/s11011-019-0389-5.
32. Lin, H., Tsai, J., Wu, L., et al., (2020). Reveals of new candidate active components in *Hemerocallis* radix and its anti-depression action of mechanism based on network pharmacology approach. *Int. J. Mol. Sci., 21*(5), 1868. doi: 10.3390/ijms21051868.
33. Fang, K., Li, H. R., Chen, X. X., et al., (2020). Quercetin alleviates LPS-induced depression-like behavior in rats via regulating BDNF-related imbalance of copine 6 and TREM1/2 in the hippocampus and PFC. *Front Pharmacol., 10*, 1544. doi: 10.3389/fphar.2019.01544.
34. Khan, K., Najmi, A. K., & Akhtar, M., (2019). A natural phenolic compound quercetin showed the usefulness by targeting inflammatory, oxidative stress markers and augment 5-HT levels in one of the animal models of depression in mice. *Drug Res., 69*(07), 392–400. doi: 10.1055/a-0748–5518.
35. Şahin, T. D., Gocmez, S. S., Duruksu, G., et al., (2019). Resveratrol and quercetin attenuate depressive-like behavior and restore impaired contractility of vas deferens in chronic stress-exposed rats: Involvement of oxidative stress and inflammation. *N-S Arch Pharmacol., 393*(5), 761–775. doi: 10.1007/s00210-019-01781-5.
36. Mehta, V., Parashar, A., & Udayabanu, M., (2017). Quercetin prevents chronic unpredictable stress induced behavioral dysfunction in mice by alleviating hippocampal oxidative and inflammatory stress. *Physiol. Behav., 171*, 69–78. doi: 10.1016/j.physbeh.2017.01.006.
37. Holzmann, I., Silva, L. M., Silva, J. A., et al., (2015). Antidepressant-like effect of quercetin in bulbectomized mice and involvement of the antioxidant defenses, and the glutamatergic and oxidonitrergic pathways. *Pharmacol. Biochem. Be., 136*, 55–63. doi: 10.1016/j.pbb.2015.07.003.
38. Dongyi, H., Jiayu, G., Xiao, Y., & Ying, L., (2021). Chinese pharmacopoeia revisited: A review of anti-depression herbal sources. *Natural Product Communications, 16*(12), 1–18.
39. Parashar, A., Mehta, V., & Udayabanu, M., (2017). Rutin alleviates chronic unpredictable stress-induced behavioral alterations and hippocampal damage in mice. *Neurosci. Lett., 656*, 65–71. doi: 10.1016/j.neulet.2017.04.058.
40. Anjomshoa, M., Boroujeni, S. N., Ghasemi, S., et al., (2020). Rutin via increase in the CA3 diameter of the hippocampus exerted anti-depressant like effect in mouse model of maternal separation stress: Possible involvement of NMDA receptors. *Behav. Neurol., 2020*, 4813616. doi: 10.1155/2020/4813616.
41. Machado, D. G., Bettio, L. E. B., Cunha, M. P., et al., (2008). Antidepressant-like effect of rutin isolated from the ethanolic extract from *Schinus molle* L. in mice: Evidence for the involvement of the serotonergic and noradrenergic systems. *Eur. J. Pharmacol., 587*(1–3), 163–168. doi: 10.1016/j.ejphar.2008.03.021.
42. Zeni, A. L. B., Camargo, A., & Dalmagro, A. P., (2017). Ferulic acid reverses depression-like behavior and oxidative stress induced by chronic corticosterone treatment in mice. *Steroids, 125*, 131–136. doi: 10.1016/j.steroids.2017.07.006.
43. Liu, Y. M., Hu, C. Y., Shen, J. D., Wu, S. H., Li, Y. C., & Yi, L. T., (2017a). Elevation of synaptic protein is associated with the antidepressant-like effects of ferulic acid in a chronic model of depression. *Physiol. Behav., 169*, 184–188. doi: 10.1016/j.physbeh.2016.12.003.

44. Moghadas, M., Edalatmanesh, M. A., & Robati, R., (2016). Histopathological analysis from gallic acid administration on hippocampal cell density, depression, and anxiety related behaviors in a trimethyltin intoxication model. *Cell J., 17*(4), 659–667.
45. Chhillar, R., & Dhingra, D., (2013). Antidepressant-like activity of gallic acid in mice subjected to unpredictable chronic mild stress. *Fund Clin. Pharmacol., 27*(4), 409–418. doi: 10.1111/j.1472-8206.2012.01040.x.
46. Nabavi, S. F., Habtemariam, S., Lorenzo, A. D., et al., (2016). Post-stroke depression modulation and *in vivo* antioxidant activity of gallic acid and its synthetic derivatives in a murine model system. *Nutrients, 8*(5), 248. doi: 10.3390/nu8050248.
47. Samad, N., Jabeen, S., Imran, I., et al., (2019). Protective effect of gallic acid against arsenic-induced anxiety-/depression- like behaviors and memory impairment in male rats. *Metab. Brain Dis., 34,* 1091–1102. doi: 10.1007/s11011-019-00432-1.
48. Can, Ö. D., Turan, N., Özkay, Ü. D., et al., (2017). Antidepressant-like effect of gallic acid: Dual involvement of serotonergic and catecholaminergic system. *Life Sci., 190,* 110–117. doi: 10.1016/j.lfs.2017.09.023.
49. Szwajgier, D., Borowiec, K., & Pustelniak, K., (2017). The neuroprotective effects of phenolic acids: Molecular mechanism of action. *Nutrients, 9*(5), 477. doi: 10.3390/nu9050477.
50. Dastmalchi, K., Damien, D. H. J., Oinonen, P. P., Darwis, Y., Laakso, I., & Hiltunen, R., (2008). Chemical composition and *in vitro* antioxidative activity of a lemon balm (*Melissa officinalis* L.) extract. *Science and Technology, 41*(3), 391–400.
51. Taiwo, A. E., Leite, F. B., Lucena, G. M., et al., (2012). Anxiolytic and antidepressant-like effects of *Melissa officinalis* (lemon balm) extract in rats: Influence of administration and gender. *The Indian Journal of Pharmacology, 44*(2), 189–192.
52. Fusar-Poli, L., Vozza, L., Gabbiadini, A., et al., (2019). Curcumin for depression: A meta-analysis. *Crit. Rev. Food Sci., 60*(15), 2643–2653. doi: 10.1080/10408398.2019.1653260.
53. Seo, H. J., Wang, S. M., Han, C., et al., (2015). Curcumin as a putative antidepressant. *Expert Rev. Neurother., 15*(3), 269–280. doi: 10.1586/14737175.2015.1008457.
54. Naqvi, F., Saleem, S., Naqvi, F., et al., (2019). Curcumin lessens unpredictable chronic mild stress-induced depression and memory deficits by modulating oxidative stress and cholinergic activity. *Pak. J. Pharm. Sci., 32*(4), 1893–1900.
55. Fan, C. Q., Song, Q. Q., Wang, P., et al., (2019). Neuroprotective effects of curcumin on IL-1 beta-induced neuronal apoptosis and depression like behaviors caused by chronic stress in rats. *Front Cell Neurosci., 12*, 516. doi: 10.3389/fncel.2018.00516.
56. Hurley, L. L., Akinfiresoye, L., Nwulia, E., et al., (2013). Antidepressant-like effects of curcumin in WKY rat model of depression is associated with an increase in hippocampal BDNF. *Behav. Brain Res., 239*, 27–30. doi: 10.1016/j.bbr.2012.10.049.
57. Liu, D., Wang, Z., Gao, Z., et al., (2014). Effects of curcumin on learning and memory deficits, BDNF, and ERK protein expression in rats exposed to chronic unpredictable stress. *Behav. Brain Res., 27*, 116–121. doi: 10.1016/j.bbr.2014.05.068.
58. Zhang, L., Xu, T., Wang, S., et al., (2012). Curcumin produces antidepressant effects via activating MAPK/ERK-dependent brain-derived neurotrophic factor expression in the amygdala of mice. *Behav. Brain Res., 235*(1), 67–72. doi: 10.1016/j.bbr.2012.07.019.

59. Zhang, L., Luo, J., Zhang, M., et al., (2014). Effects of curcumin on chronic, unpredictable, mild, stress-induced depressive-like behavior and structural plasticity in the lateral amygdala of rats. *Int. J. Neuropsychoph., 17*(05), 793–806. doi: 10.1017/s1461145713001661.
60. Chang, X., Wang, L., Li, J., et al., (2016). Analysis of anti-depressant potential of curcumin against depression induced male albino Wistar rats. *Brain Res., 1642*, 219–225. doi: 10.1016/j.brainres.2016.03.010.
61. Madiha, S., & Haider, S., (2019). Curcumin restores rotenone induced depressive-like symptoms in animal model of neurotoxicity: Assessment by social interaction test and sucrose preference test. *Metab. Brain Dis., 34*(1), 297–308. doi: 10.1007/s11011-018-0352-x.
62. Xu, Y., Ku, B. S., Yao, H. Y., et al., (2005). Antidepressant effects of curcumin in the forced swim test and olfactory bulbectomy models of depression in rats. *Pharmacol. Biochem. Be., 82*(1), 200–206. doi: 10.1016/j.pbb.2005.08.009.
63. Choi, G. Y., Kim, H. B., Hwang, E. S., et al., (2017). Curcumin alters neural plasticity and viability of intact hippocampal circuits and attenuates behavioral despair and cox-2 expression in chronically stressed rats. *Mediat. Inflamm., 2017*, 6280925. doi: 10.1155/2017/6280925.
64. Rinwa, P., Kumar, A., & Garg, S., (2013). Suppression of neuroinflammatory and apoptotic signaling cascade by curcumin alone and in combination with piperine in rat model of olfactory bulbectomy induced depression. *PloS One, 8*(4), e61052. doi: 10.1371/journal.pone.0061052.
65. Jiang, H., Wang, Z., Wang, Y., et al., (2013). Antidepressant-like effects of curcumin in chronic mild stress of rats: Involvement of its anti-inflammatory action. *Prog. Neuro-Psychoph., 47*, 33–39. doi: 10.1016/j.pnpbp.2013.07.009.
66. Wang, Z., Zhang, Q., Yuan, L., et al., (2014). The effects of curcumin on depressive-like behavior in mice after lipopolysaccharide administration. *Behav. Brain Res., 274*, 282–290. doi: 10.1016/j.bbr.2014.08.018.
67. Zhang, W., Guo, Y., Han, W., et al., (2019). Curcumin relieves depressive like behaviors via inhibition of the NLRP3 inflammasome and kynurenine pathway in rats suffering from chronic unpredictable mild stress. *Int. Immunopharmacol., 67*, 138–144. doi: 10.1016/j.intimp.2018.12.012.
68. Lin, Z., Shi, L., Lu, J., et al., (2013). Effects of curcumin on glucose metabolism in the brains of rats subjected to chronic unpredictable stress: An 18 F-FDG micro-PET study. *BMC Complem Altern. M., 13*(1), 202. doi: 10.1186/1472-6882-13-202.
69. Lin, T. Y., Lu, C. W., Wang, C. C., et al., (2011). Curcumin inhibits glutamate release in nerve terminals from rat prefrontal cortex: Possible relevance to its antidepressant mechanism. *Prog. Neuro-Psychoph., 35*(7), 1785–1793. doi: 10.1016/j.pnpbp.2011.06.012.
70. Khare, P., Chaudhary, S., Singh, L., Yadav, G., & Verma, S., (2014). Evaluation of nootropic activity of *Cressa cretica* in scopolamine induced memory impairment in mice. *International Journal of Pharmacology and Toxicology, 2*(2), 24–29. doi: 10.14419/ijpt.v2i2. 2004.
71. Wang, Y., Han, T., Zhu, Y., Zheng, C. J., Ming, Q. L., Rahman, K., & Qin, L. P., (2010). Antidepressant properties of bioactive fractions from the extract of *Crocus sativus* L. *Journal of Natural Medicines, 64*(1), 24–30. doi: 10.1007/s11418-009-0360-6.

72. Rios, J., Recio, M., Giner, R., & Manez, S., (1996). An update review of saffron and its active constituents. *Phytother. Res., 10*, 189–193.
73. Karimi, G. R., Hosseinzadeh, H., & Khalegh, P. P., (2001). Study of antidepressant effect of aqueous and ethanolic extract of *Crocus sativus* in mice. *Iran J. Basic Med. Sci., 4*, 11–15.
74. Hosseinzadeh, H., Karimi, G., & Niapoor, M., (2003). Antidepressant effect of *Crocus sativus* L. stigma extracts and their constituents, crocin and safranal, in mice. In: *I International Symposium on Saffron Biology and Biotechnology* (Vol. 650, pp. 435–445).
75. Hosseinzadeh, H., & Jahanian, Z., (2010). Effect of *Crocus sativus* L (saffron) stigma and its constituents, crocin and safranal, on morphine withdrawal syndrome in mice. *Phytother. Res., 24*, 726–730.
76. Verma, S. K., & Bordia, A., (1998). Antioxidant property of saffron in man. *Indian J. Med. Sci., 52*, 205–207.
77. Georgiadou, G., Tarantilis, P. A., & Pitsikas, N., (2012). Effects of the active constituents of *Crocus sativus* L., crocins, in an animal model of obsessive compulsive disorder. *Neurosci. Lett., 528*, 27–30.
78. Pitsikas, N., Boultadakis, A., Georgiadou, G., Tarantilis, P. A., & Sakellaridis, N., (2008). Effects of the active constituents of *Crocus sativus* L., crocins, in an animal model of anxiety. *Phytomedicine, 15*, 1135–1139.
79. Hosseinzadeh, H., & Noraei, N. B., (2009). Anxiolytic and hypnotic effect of *Crocus sativus* aqueous extract and its constituents, crocin and safranal, in mice. *Phytother. Res., 23*, 768–774.
80. Talaei, A., Hassanpour, M. M., Sajadi, T. S. A., & Mohajeri, S. A., (2015). Crocin, the main active saffron constituent, as an adjunctive treatment in major depressive disorder: A randomized, double-blind, placebo-controlled, pilot clinical trial. *J. Affect. Disord., 174*, 51–56.
81. Akhondzadeh, S., Fallah-Pour, H., Afkham, K., Jamshidi, A. H., & Khalighi-Cigaroudi, F., (2004). Comparison of *Crocus sativus* L. and imipramine in the treatment of mild to moderate depression: A pilot double-blind randomized trial [ISRCTN45683816]. *BMC Complement. Altern. Med., 4*, 12. Available from: https://doi.org/10.1186/1472-6882-4-12.
82. Kianbakht, S., & Ghazavi, A., (2005). Evaluation of immunological and hematological effects of saffron in men. *Ethnopharmacologia, 36*, 78–83.
83. Amin, B., Nakhsaz, A., & Hosseinzadeh, H., (2015). Evaluation of the antidepressant-like effects of acute and sub-acute administration of crocin and crocetin in mice. *Avicenna J. Phytomedicine, 5*(5), 458–468. doi: 10.22038/ajp.2015.4619.
84. Xiao, Q., Xiong, Z., Yu, C., et al., (2019). Antidepressant activity of crocin-I is associated with amelioration of neuroinflammation and attenuates oxidative damage induced by corticosterone in mice. *Physiol. Behav., 212*, 112699. doi: 10.1016/j.physbeh.2019.112699.
85. Hassani, F. V., Naseri, V., Razavi, B., et al., (2014). Antidepressant effects of crocin and its effects on transcript and protein levels of CREB, BDNF, and VGF in rat hippocampus. *Daru., 22*(1), 16. doi: 10.1186/2008-2231-22-16.

86. Tang, J., Lu, L., Wang, Q., et al., (2020). Crocin reverses depression-like behavior in Parkinson disease mice via VTA-mPFC pathway. *Mol. Neurobiol., 57*(7), 3158–3170. doi: 10.1007/s12035-020-01941-2.
87. Xie, Y., He, Q., Chen, H., et al., (2019). Crocin ameliorates chronic obstructive pulmonary disease-induced depression via PI3K/Akt mediated suppression of inflammation. *Eur. J. Pharmacol., 862*, 172640. doi: 10.1016/j.ejphar.2019.172640.
88. Zhang, L., Previn, R., Lu, L., et al., (2018). Crocin, a natural product attenuates lipopolysaccharide-induced anxiety and depressive-like behaviors through suppressing NF-kB and NLRP3 signaling pathway. *Brain Res Bull., 142*, 352-359. doi: 10.1016/j.brainresbull.2018.08.021.
89. Wu, R., Xiao, D., Shan, X., et al., (2020). Rapid and prolonged antidepressant-like effect of crocin is associated with GHSR mediated hippocampal plasticity-related proteins in mice exposed to prenatal stress. *ACS Chem. Neurosci., 11*(8), 1159-1170. doi: 10.1021/acschemneuro.0c00022.
90. Farkhondeh, T., Samarghandian, S., Samini, F., et al., (2018). Protective effects of crocetin on depression-like behavior induced by immobilization in rat. *CNS Neurol Disord-Dr., 17*(5), 361-369. doi:10.2174/1871527317666180515120212.
91. Fu, S., Wang, J., Hao, C., Dang, H., & Jiang, S., (2019). Tetramethylpyrazine ameliorates depression by inhibiting TLR4-NLRP3 inflammasome signal pathway in mice. *Psychopharmacology (Berl), 236*, 2173–2185. doi:10.1007/s00213-019-05210-6.
92. Ma, J. C., Zhang, H. L., Huang, H. P., Ma, Z. L., Chen, S. F., Qiu, Z. K., et al., (2021). Antidepressant-like effects of Z-ligustilide on chronic unpredictable mild stress-induced depression in rats. *Exp. Ther. Med., 22*, 677. doi:10.3892/etm.2021.10109.
93. Zhu, X., Jing, L., Chen, C., Shao, M., Fan, Q., Diao, J., et al., (2015). Danzhi Xiaoyao San ameliorates depressive-like behavior by shifting toward serotonin via the downregulation of hippocampal indoleamine 2,3-dioxygenase. *J. Ethnopharmacol., 160*, 86–93. doi: 10.1016/j.jep.2014.11.031.3074237396.

# Index

## B

## G

## H

## O

## T

## U



## V

## X



## Y



## Z